John S. Nkoma
and Pushpendra K. Jain

INTRODUCTION TO OPTICS

Geometrical, Physical and Quantum

MKUKI NA NYOTA
DAR — ES — SALAAM

PUBLISHED BY

Mkuki na Nyota Publishers Ltd
P. O. Box 4246
Dar es Salaam, Tanzania
www.mkukinanyota.com

First published 2003 by Bay Publishers, Gaborone
Second Edition 2019 by Mkuki na Nyota, Dar es Salaam

© John S. Nkoma and Pushpendra K. Jain, 2019

ISBN 978-9987-08-371-8

Typesetting by John S. Nkoma

Visit www.mkukinanyota.com to read more about and to purchase any of Mkuki na Nyota books. You will also find featured authors, interviews and news about other publisher/author events. Sign up for our e-newsletters for updates on new releases and other announcements.

Distributed world wide outside Africa by African Books Collective.
www.africanbookscollective.com

Contents

CONTENTS

CONTENTS

The Authors

Professor J S Nkoma obtained his BSc in Physics and Mathematics from the University of Dar es Salaam, Tanzania, and later his MSc and PhD in Physics from the University of Essex, UK. Prof. Nkoma has over 30 years experience of university teaching and research in Tanzania, Botswana, Italy and UK, and has been external examiner to several universities. His research interests are in condensed matter physics, materials science and ICT Regulation in telecommunications/internet, broadcasting and postal communications. He is a Fellow of the African Academy of Sciences. He also served as the Director General/CEO of the Tanzania Communications Regulatory Authority (TCRA) during 2004 to 2015. He recently published the title: Introduction to Basic Concepts for Engineers and Scientists: Electromagnetic, Quantum, Statistical and Relativistic Concepts, John. S. Nkoma, Mkuki na Nyota Publishers (2018). He can be contacted at his emailaddress: jsnkoma@hotmail.com

Professor P K Jain holds a PhD degree in physics from the University of Connecticut, USA, and MSc and BSc degrees from India. He is a Chartered Physicist (CPhys) and a Fellow of the Institute of Physics (FInstP), UK. Prof. Jain has over 50 years of teaching and research experience in India, USA, Zambia including past 31 years in Botswana. He has been external examiner at several universities in Africa. Prof. Jain has made over 150 contributions to research in materials science, renewable energy, and general physics, has lectured internationally, and has received several awards and honours. Prof. Jain has been the UB Coordinator (2008-2016) of the African Materials Science and Engineering Network: AMSEN a Carnegie-IAS-RISE network of six African universities with the aim to train manpower at MSc and PhD degree levels, to build supervisory capacity and to develop research infrastructure in materials science and engineering in member universities. He can be contacted at his email address: jainpk@mopipi.ub.bw

Professors J S Nkoma and P K Jain have co-authored these two books:
1. Introduction to Optics: Geometrical, Physical and Quantum, J S Nkoma and P K Jain, Bay Publishers, Gaborone ISBN 99912-511-6-2 (2003) and Second Edition, Mkuki na Nyota Publishers, Dar es Salaam (2019), and
2. Introduction to Mechanics: Kinematics, Newtonian and Lagrangian, P K Jain and J S Nkoma, Bay Publishers, Gaborone ISBN 99912-561-4-8 (2004) and Second Edition, Introduction to Classical Mechanics: Kinematics, Newtonian and Lagrangian, Mkuki na Nyota Publishers, Dar es Salaam (2019).

Preface to the Second Edition

The first edition of our book *Introduction to Optics: Geometrical, Physical and Quantum* was published fifteen years ago by Bay Publishers in Gaborone, Botswana. The second edition, is published by Mkuki na Nyota Publishers in Dar es Salaam, Tanzania. The Foreword, Preface and Content of the first edition are still relevant and are reproduced herein in this second edition.

Why the second edition? First, we have updated the contents, added Appendices for some useful constants and mathematical relations, and fixed a few typos which were present in the first edition. Secondly, we have added a completely new feature by having a chapter summary at the end of each chapter so as to enhance learning. Thirdly, the current publisher, Mkuki na Nyota has ably redesigned the book and produced this second edition with vision, excellence and professionalism. Fourthly, with our objective of contributing to the training of scientists and engineers, we have now with Mkuki na Nyota Publishers three books: 1. Introduction to Basic Concepts for Engineers and Scientists: Electromagnetic, Quantum, Statistical and Relativistic Concepts by John. S. Nkoma, Mkuki na Nyota Publishers (2018), 2. Introduction to Classical Mechanics: Kinematics, Newtonian and Lagrangian by Pushpendra K Jain and John S Nkoma, Second edition (2018) and 3. Introduction to Optics: Geometrical, Physical and Quantum by John S Nkoma and Pushpendra K Jain, Second edition (2018), which can meet that objective, and the trained graduates in turn will contribute to the industrial and economic development in the region.

In this second edition, the authors still acknowledge colleagues mentioned in the preface of the first edition as well as our many students over the years. We also thank staff of Mkuki na Nyota Publishers for several discussions during manuscript preparation and for finally producing the book with an excellent quality. Patience and understanding of our families and friends is gratefully acknowledged.

John S. Nkoma
Professor of Physics and
Ex-Director General/CEO, TCRA
Mbezi Beach
Dar es Salaam, Tanzania

October 2018

Pushpendra K. Jain
Professor of Physics
Department of Physics
University of Botswana
Gaborone, Botswana

October 2018

Foreword to the First Edition

Most countries in Africa lack adequate trained indigenous manpower in science and technology who are key role players in industrial and economic development in any society. Some countries, even after 40 or more years of independence from colonialism, still lack the very basic need of trained science teachers. This can mainly be attributed to several reasons including the high cost of training in the areas of science and technology, the cost of text books and their availability. Undoubtedly, plenty of text books in every area of science and technology are available on the world market, but they pose a number of limitations. In view of the weak currencies, these books invariably turn out to be exorbitantly costly, and foreign exchange scarcity leads to delays if not difficulty in the procurement. Therefore, the development of indigenous quality text books by mature and experienced African academics and professionals for the African market is as important to the much awaited technological revolution on the continent as the training itself is. The book *Introduction to Optics: Geometrical, Physical and Quantum* for undergraduate degree students by two senior and highly experienced academics in Africa, Professors J S Nkoma and P K Jain is a most welcome contribution to Physics education. Physics is the fundamental science required by all disciplines of science and technology in so much so that it can be termed the Mother of all Sciences. Not only is the book locally produced at an affordable cost to readers, it has some very special features to meet the needs of African readership. The book takes into conscience the background of the entry students who come with varied preparation: it is written in simple to understand language even for students who may initially have some language barrier, diagrams are simple, well labeled and easy to comprehend, and concepts are elaborately discussed. However, I must add that physics being a universal science, so is the book which shall be found useful by students globally.

I commend Professors Nkoma and Jain for their concern and contribution to African scientific manpower training. I hope more academics like them will take the initiative to put their experience and expertise in print for the benefit of the African masses to make training in science and technology more accessible to them.

Juma Shabani
Professor of Mathematical Physics and
Director, UNESCO Harare Cluster Office
July 2003

Preface to the First Edition

Contemporary physics programmes are under increasing pressure to provide a balance between coverage of several traditional branches of physics such as optics, mechanics, electromagnetism, thermodynamics, quantum physics, relativity and statitistical physics on the one hand, and to expose students to emerging research areas on the other hand. For this reason, it is important to provide an indepth introduction to some branches of physics, such as optics, to students who may not become professional physicists but will need physics in their chosen professions. The purpose of this book, "Introduction to Optics: Geometrical, Physical and Quantum" is to introduce University undergraduates to the fascinating world of the science of light. In some Universities, Optics may be offered as semester courses while in others it may be offered as modules within General Physics courses in the degree programme. The book meets the needs of both approaches.

Optics is the study of light, and there are three major branches: Geometrical optics, Physical optics and Quantum optics. The nature of light is discussed in chapter 1. Geometrical optics is covered in chapters 2, 3, 4, and 5. The Laws of geometrical optics are treated in chapter 2. Spherical mirrors and thin spherical lenses are covered in chapters 3 and 4 respectively, and chapter 5 gives the applications of the laws of geometrical optics to optical instrumentation. Physical optics is treated in chapters 6, 7 and 8, with three important properties of light: interference, diffraction and polarization discussed respectively. Quantum optics is introduced in chapter 9, which lays a foundation for advanced courses in applied quantum optics.

Towards the end of this book, there is a comprehensive bibliography of books and a few journal articles which students are advised to consult. The bibliography, arranged alphabetically, includes both recent references and a number of old editions of books which are still considered to be authoritative writings.

It is well recognized that the language of physics is universal, and the book is suited to students globally. However, we recognise certain peculiarities in Africa. It is with this in mind, that the book is written to meet the specific needs of students in African Universities, who come with widely varied background and preparation. On one side, there are students from well equipped and well staffed schools and on the other side there are students from not so well equipped and staffed schools, generally in remote rural settings. These two groups of students attending the same courses together at one of the very few (sometimes only one) Universities in the country have very different needs. The well prepared students need a bit of challenge in their work, while the others need to be taught in fair detail to maintain their interest, and to ensure that by the end of the course both groups have reached near equal levels of competence in the subject. To meet these needs, this book on optics has detailed discussions and explanations of difficult to grasp topics with the help of simple but clearly drawn and labeled diagrams. The discussions and important conclusions are presented pointwise, and important key words, definitions, laws etc are highlighted. A number of tasks similar to those discussed are assigned as exercises for students to present them with a bit of challenge. There are a large number of problems and exercises at the end of each chapter. The book distills over 60 years of combined teaching experience of two authors in Botswana, India, Tanzania, UK, USA and Zambia, and the experience of having been external examiners at many Universities in Southern and Eastern Africa, as well as serving in other capacities associated with Science and Technology.

Acknowledgments

The authors would like to express their sincere gratitude to a number of colleagues: Professor E M Lungu, Dr. P V C Luhanga, Dr J Prakash, Dr L K Sharma, Dr. T S Verma and Dr D P Winkoun for carefully reading through the manuscript and for making many valuable suggestions. We also thank our publisher for having produced the book so well and for frequent discussions on manuscript preparation. Patience and understanding of our families for having stayed away from them for extended hours and over weekends during the course of this work are gratefully acknowledged. Lastly, but not least, we thank our over 25,000 students over the years who gave us the opportunity to sharpen our skills, and our understanding of the physics reflected in the book.

J S Nkoma and P K Jain
Department of Physics
University of Botswana
Gaborone
July 2003

Chapter 1

Propagation and Nature of Light

1.1 Introduction

Light is one of the basic essentials of life. Light enables human beings and animal life in general to see other objects. Plants convert light energy from the sun into chemical energy through the process of photosynthesis. Minerals are known to interact with light and produce beautiful colours. Cosmic bodies such as the Sun in our solar system are sources of light. Other sources include fire, electronic transitions in atoms. In technology, light is important in transmission and reception of signals. These are some of the reasons that make a study of light to be so important.

The study of light is known as "Optics", and there are three major branches: Geometrical optics, Physical optics and Quantum optics, with various topics as summarised below:

- *Geometrical optics*
  Reflection
  Refraction
  Dispersion

- *Physical optics*
  Interference
  Diffraction
  Polarisation

- *Quantum optics*
  Absorption
  Emission
  Scattering

The topics under geometrical optics are covered in chapters 2, 3, 4 and chapter 5 presents their applications to optical instrumentation. The topics under the second branch, physical optics, are treated in chapters 6, 7 and 8. The topics under the third branch, quantum optics, are introduced in chapter 9.

1.2   Rectilinear propagation of light

Light is observed to travel in straight lines. This well known principle is what is referred to as *rectilinear propagation of light*. One of the everyday demonstrations of the fact that light travels in a straight line is the observation that objects cast shadows in presence of light. Other demonstrations are image formation, for example by a pinhole camera and eclipses of celestial bodies as discussed below. The path of propagation of light is shown by a straight line with an arrow head indicating the direction of propagation, know as a "ray" of light.

1.2.1   Image formation by a pinhole camera

The pinhole camera is illustrated in Figure 1.1. Rays of light strike an object, AB,pass through a screen with a small pinhole and an image, A'B', is formed on the second screen (or film). The observed image is inverted.

The image formation can be explained as follows. Consider two points, A and B, at the top and bottom of the object respectively. The ray from A passes through the pinhole to the point A' in a straight line, while the ray from B will arrive at B' in a straight line as well. The image $B'A'$ is inverted, that is its orientation is opposite to that of the object AB. The size of the image $B'A'$ depends on its distance from the pinhole as well as the distance of the object from the pinhole. When the image screen is moved away from the pinhole, the image is larger, while when it is moved closer to the pinhole the image is smaller. Likewise, the image becomes larger as the object is moved towards the pinhole, and becomes smaller as the object is moved away from the pinhole.

The student should attempt to explain the following. How are the angles subtended by the object and by the image at the pinhole related? How does the size of the image change in terms of these angles? The angle subtended by an object (image) at a point is known as the *angular size* of the object (image) seen by that point.

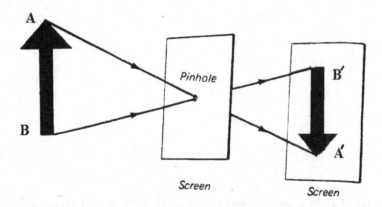

Figure 1.1: A pinhole camera demonstrating the principle of rectilinear propagation of light.

1.2.2 Eclipses

An eclipse is an astronomical event in which three celestial bodies are aligned, with the luminous body at one end and the shadow of the body in the middle is cast on to the body at the far end. This can be illustrated by three bodies: the Sun, Earth and the Moon. Let us discuss two eclipses which can be observed from the earth:

- Eclipse of the Moon

- Eclipse of the Sun

As is well known, the Earth revolves around the Sun, and in turn, the Moon revolves around the Earth. At certain instances, these three bodies are aligned such that the shadow of the Earth falls on to the Moon, as illustrated in Figure 1.2. This is called the *eclipse of the Moon* (also known as the Lunar eclipse). In Figure 1.2, two regions of interest can be identified: the *Umbra* with a total shadow and the *Penumbra* with a partial shadow. The Lunar eclipse is an evidence that light travels in a straight line and can safely be observed with bare eyes, field glasses or a small telescope. Figure 1.2 is not to scale. The sizes of the Earth, Sun and Moon and their relative distances are given in Table 1.1.

Table 1.1: Average distances involving the Sun, Earth and the Moon.

Sun-Earth distance	1.496×10^{11} m
Radius of the Sun	6.960×10^{8} m
Earth-Moon distance	3.844×10^{8} m
Radius of the Earth	6.378×10^{6} m
Radius of the Moon	1.740×10^{6} m

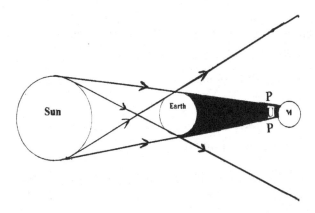

Figure 1.2 (not to scale): Eclipse of the Moon (M). U (*Umbra*) denotes total shadow and P (*Penumbra*) denotes partial shadow

During the *eclipse of the Sun* (also known as the Solar eclipse), the shadow of the Moon due to light from the Sun falls on the Earth, as illustrated in Figure 1.3 (not to scale). Two regions of interest can be identified: the *Umbra* with a total shadow and the *Penumbra* with a partial shadow. The totality of the solar eclipse can only be observed along a narrow belt on the Earth on relatively rare occassions. The solar eclipse *must not* be viewed with the bare eyes or directly with a telescope. Why? Explain with a suitable diagram how the umbral shadow is affected when the earth-moon distance is large. Will the total solar eclipse be seen on earth on such occassions?

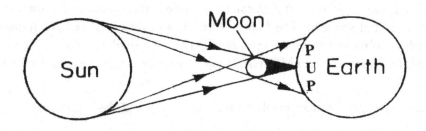

Figure 1.3 (not to scale): Eclipse of the Sun. U (*Umbra*) denotes total shadow and P (*Penumbra*) denotes partial shadow.

1.3   Speed of light

Although light travels with an extremely high speed, the speed is *finite* and not infinite. The value of the speed of light, c, in free space (vacuum) is given by

$$
\begin{aligned}
c &= 2.997925 \times 10^8 \text{ m.s}^{-1} \\
&\approx 3.0 \times 10^8 \text{ m.s}^{-1} \\
&\approx 300000 \text{ km.s}^{-1} \\
&\approx 186000 \text{ mi.s}^{-1}
\end{aligned}
\tag{1.1}
$$

in different units, although SI units will normally be used in this book. According to the theory of relativity, the speed of light is the highest possible speed in nature. To date several methods of determining the velocity of light have been designed, and we describe the following two methods because of their historic importance.

- Fizeau's method

- Michelsons's method

1.3.1 Fizeau's method

The first attempt to measure the speed of light was made by Armand H L Fizeau, a French physicist, in 1849. He used a method which is referred to as Fizeau's method, and is illustrated in Figure 1.4. The method consists of a light source S, toothed wheel and a mirror M. L_1, L_2 and L_3 are lenses. Suppose d is the distance between the toothed wheel and the mirror. If t is the transit time for the round trip of light between the toothed wheel and the mirror, then

$$\text{Speed of light, } c = \frac{2d}{t}$$

In the experiment, the toothed wheel is rotated at an optimal lowest speed such that the observer sees the maximum intensity of the light reflected by the mirror. In that case, time t is just equal to the time taken by the wheel to rotate from one gap to the next.

If the wheel has n teeth and n gaps and rotates at N revolutions per second, then $t = 1/(Nn)$ (Show this!). Hence, the speed of light, c, using Fizeau's method is given as

$$c = \frac{2d}{t} = 2dnN$$

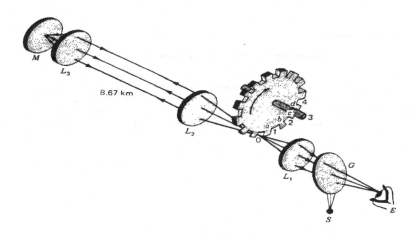

Figure 1.4: Experimental arrangement used in the Fizeau's method for the determination of the speed of light in air.

Fizeau's experiment used the following values: $d = 8.67$km, $n = 720$, $N = 25$rev/s, and obtained $c = 312000$ km/s$= 3.12 \times 10^8$ m/s, a value which is within experimental error to that obtained from today's more sophisticated experiments.

1.3.2 Michelson's method

Another major attempt to measure the speed of light was made by A A Michelson, an American physicist, and coworkers in 1926. They used a method which is referred to as Michelson's method, and illustrated in Figure 1.5. The method consists of a light source, an 8 sided rotating mirror, fixed mirrors and a detector. Suppose d is the distance between the 8 sided rotating mirror and the fixed mirrors. If t is the transit time for the round trip of light between the 8 sided rotating mirror and the fixed mirrors, then

$$\text{Speed of light, } c = \frac{2d}{t}$$

In the experiment, the 8 sided mirror is rotated at an optimal lowest speed such that the detector records the maximum intensity of the light reflected by the mirror. If the 8 sided mirror rotates at N revolutions per second, in time t there is $1/8$ revolution, and hence $N = 1/8t$, which gives the speed of light, c, using Michelson's method as

$$c = \frac{2d}{t} = 16dN$$

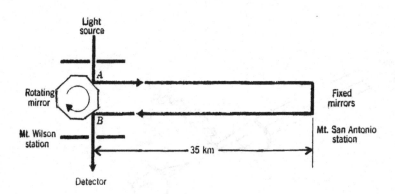

Figure 1.5: Experimental arrangement used in the Michelson's method for the determination of the speed of light in air.

Michelson's experiment used the following values: $d = 35.5$km, $N = 528$ rev/s, and obtained $c = 299796$ km/s$= 2.99796 \times 10^8$ m/s, a value which is within experimental error to that obtained using more sophisticated techniques in the modern era.

1.4   Nature of Light: Wave or particle?

Light is commonly observed in everyday life. However, the question is what is light? Is light made up of waves? Is light made up of particles? Is light a wave or particle or both? The answer to these questions is provided through experimentation. In sections 1.4.1 and 1.4.2 below , the wave nature of light is discussed, while in section 1.4.3 the particle nature is discussed. This means that light can behave as either a wave, or as a particle, and hence light shows wave-particle duality as we shall see in section 1.4.4.

1.4.1 Wave nature of light and the Electromagnetic spectrum

The wave nature of light is best illustrated by waves comprising the electromagnetic spectrum. The electromagnetic spectrum is a family of a wide range of waves which differ in terms of their wavelength or frequency. These ranges, in order of increasing frequency (or decreasing wavelength) include Radio waves, Microwaves, Infrared, Visible, Ultraviolet, X-rays and Gamma-rays. The classification of the various ranges does not have sharp boundaries. The electromagnetic spectrum is illustrated in Figure 1.6.

The relation between the speed of light in vacuum, c, wavelength, λ and frequency, ν is

$$c = \lambda \nu \tag{1.2}$$

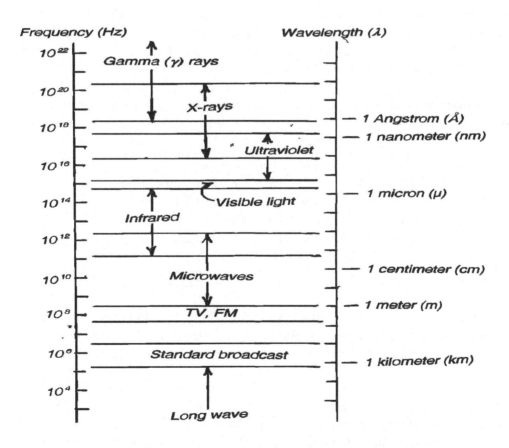

Figure 1.6: The electromagnetic spectrum.

The various types of waves comprising the electromagnetic spectrum are described below:

- *Radio waves:* These waves have wavelengths ranging from a few kilometers to 0.3 m, and the corresponding frequency range is from a few Hz to 10^9 Hz. These waves are generated by electronic devices and are utilised in Radio and Television broadcasting systems.

- *Microwaves:* These waves have wavelengths ranging from 0.3 m down to 10^{-3} m and the corresponding frequency range is from 10^9 Hz to 3×10^{11} Hz. These waves are used in radar and other communication devices, and in microwave ovens.

- *Infrared:* These waves have wavelengths ranging from 10^{-3} m down to 7.8×10^{-7} m (or 7800Å), and the corresponding frequency range is from 3×10^{11} Hz to 4×10^{14} Hz. These waves are generated by molecules and hot bodies and have many applications in astronomy, industry and medicine.

- *Visible:* These waves are the ones to which the retina in our eyes is sensitive, and the wavelength range is from 7.8×10^{-7} m (or 7800Å) down to 3.8×10^{-7} m (or 3800Å), and the corresponding frequency range is from 4×10^{14} Hz to 8×10^{14} Hz. Visible light is produced by atoms and molecules as a result of electronic transitions. The different sensations that light produces on the eye is what is referred to as *colours* in everyday language. In Table 1.2, the wavelength and frequency ranges are shown for various colours.

Table 1.2: Ranges of wavelengths and frequencies in the visible portion of the electromagnetic spectrum.

Colour	Wavelength, λ (m)	Frequency, ν (Hz)
Violet	3.99 to 4.44×10^{-7}	7.69 to 6.59×10^{14}
Blue	4.55 to 4.92×10^{-7}	6.59 to 6.10×10^{14}
Green	4.92 to 5.77×10^{-7}	6.10 to 5.20×10^{14}
Yellow	5.77 to 5.97×10^{-7}	5.20 to 5.03×10^{14}
Orange	5.97 to 6.22×10^{-7}	5.03 to 4.82×10^{14}
Red	6.22 to 7.80×10^{-7}	4.82 to 3.84×10^{14}

- *Ultraviolet:* These waves have wavelengths ranging from 3.8×10^{-7} m down to 6×10^{-10} m, and the corresponding frequency range is from 8×10^{14} Hz to 3×10^{17} Hz. Overexposure to these rays from the sun is known to cause skin cancer and are also responsible for causing sun-burn.

- *X-rays:* These waves have wavelengths ranging from 10^{-9} m down to 6×10^{-12} m, and the corresponding frequency range is from 3×10^{17} Hz to 5×10^{19} Hz. These rays are used in hospitals for medical diagnostics. Overexposure to these rays is not recommended since they can cause permanent damage of body tissue.

- *Gamma-rays:* These waves have wavelengths ranging from 10^{-10} m down to 10^{-14} m, and the corresponding frequency range is from 3×10^{18} Hz to more than 3×10^{22} Hz. Gamma-rays are nuclear in origin and are produced by many radioactive substances. These rays are used for the treatment of certain types of cancers.

There are several experiments which demonstrate the wave nature of Electromagnetic waves such as Reflection (see chapters 2 and 3), Refraction (see chapters 2 and 4), Dispersion (see chapter 2), Interference (see chapter 6), Diffraction (see chapter 7) and Polarisation (see chapter 8).

1.4.2 Wave nature of light and Electromagnetic theory

According to electromagnetic theory, light is composed of the electric field $\vec{E}$ and the magnetic field $\vec{B}$. The electric field $\vec{E}$ and magnetic field $\vec{B}$ are perpendicular (*orthogonal*) to each other and are perpendicular (*transverse wave*) to the direction of propagation, as illustrated in Figure 1.7.

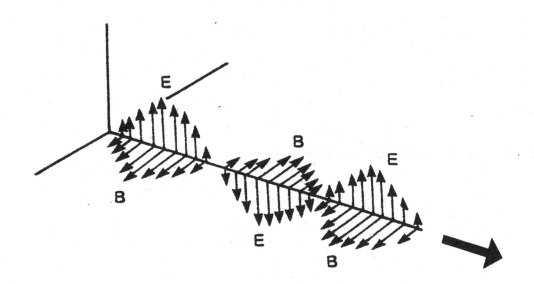

Figure 1.7: The electric field $\vec{E}$ and magnetic field $\vec{B}$ are perpendicular to each other and are perpendicular to the direction of propagation of the electromagnetic wave.

The electric field $\vec{E}$ and magnetic field $\vec{B}$ satisfy Maxwell's equations which are given below.

$$\nabla \cdot \vec{E} = \frac{\rho}{\epsilon_0} \tag{1.3}$$

$$\nabla \cdot \vec{B} = 0 \tag{1.4}$$

$$\nabla \wedge \vec{E} = -\frac{\partial \vec{B}}{\partial t} \tag{1.5}$$

$$\nabla \wedge \vec{B} = \mu_0 \vec{j} + \mu_0 \frac{\partial}{\partial t}[\epsilon_0 \epsilon(\omega)\vec{E}] \tag{1.6}$$

where

$$\rho \quad \text{is} \quad \text{the charge density,} \tag{1.7}$$

$$\epsilon_0 \quad = \quad 8.854 \times 10^{-9} Fm^{-1} \text{is the permittivity of free space,} \tag{1.8}$$

$$\mu_0 \quad = \quad 4\pi \times 10^{-7} Hm^{-1} \text{is the permeability of free space,} \tag{1.9}$$

$$\vec{j} \quad \text{is} \quad \text{the current density vector, and} \tag{1.10}$$

$$\epsilon(\omega) \quad \text{is} \quad \text{the dielectric function} \tag{1.11}$$

In free space, $\rho = 0$ and $j = 0$, $\epsilon(\omega) = 1$, Maxwell's equations reduce to

$$\nabla \cdot \vec{E} \quad = \quad 0 \tag{1.12}$$

$$\nabla \cdot \vec{B} \quad = \quad 0 \tag{1.13}$$

$$\nabla \wedge \vec{E} \quad = \quad -\frac{\partial \vec{B}}{\partial t} \tag{1.14}$$

$$\nabla \wedge \vec{B} \quad = \quad \mu_0 \epsilon_0 \frac{\partial}{\partial t} \vec{E} \tag{1.15}$$

Taking the curl of equation (1.14) and using (1.15), one obtains

$$\nabla \wedge (\nabla \wedge \vec{E}) = -\mu_0 \epsilon_0 \frac{\partial^2}{\partial t^2} \vec{E} \tag{1.16}$$

Using the identity

$$\nabla \wedge (\nabla \wedge \vec{E}) = \nabla(\nabla.\vec{E}) - \nabla^2 \vec{E} \tag{1.17}$$

and using the expression for the velocity of light, c, given by

$$c = \frac{1}{\sqrt{\mu_0 \epsilon_0}}$$

and equation (1.12), equation (1.16) leads to the wave equation

$$\nabla^2 \vec{E} = \frac{1}{c^2} \frac{\partial^2 \vec{E}}{\partial t^2} \tag{1.18}$$

A similar equation for $\vec{B}$ can be obtained, and this is left as an exercise for the student.

The one dimensional version of equation 1.18, say along x-axis is

$$\frac{\partial^2 \vec{E}}{\partial x^2} = \frac{1}{c^2} \frac{\partial^2 \vec{E}}{\partial t^2}$$

which has a solution of a wave propagating along the x - axis given by any of the following forms:

$$E = E_0 e^{i(kx - \omega t)} \tag{1.19}$$

or

$$E = E_0 \sin(kx - \omega t)$$

or

$$E = E_0 \cos(kx - \omega t)$$

which is a plane wave of amplitude E_0 with wave vector $k(= 2\pi/\lambda)$ and angular frequency $\omega(= 2\pi\nu)$. Similar equations for the magnetic field vector $\vec{B}$ can be obtained, which is left as an exercise for the student.

1.4.3 Particle nature of light and Quantum theory

The particle nature of light can be demonstrated by several experimental observations, for example: the radiation spectrum, the photoelectric effect, X-ray production and the Compton effect. Each of these observation is described briefly below.

The radiation spectrum

Experimental observation of the radiation spectrum show that:

(i) Hot solids, for example, heated metals, heated coal, electric bulb filament etc emit radiation. As the temperature rises, the dominant frequencies increase (or wavelength decreases). A typical example is a heated object becomes red hot and then bluish white on increase in temperature.

(ii) The total power emitted by a hot body is proportional to T^4, where T is absolute temperature of the hot body.

(iii) The shape of the radiation spectrum is such that it increases at low frequencies and decreases at high frequencies, with a peak in between, as illustrated in Figure 1.8.

The above observations of the radiation spectrum can not be accounted for by classical physics. Max Planck in 1900 was able to explain the radiation spectrum by developing the following model:

(i) The atoms of the walls of a hot body or a cavity emit radiation in bundles of light known as quanta, of energy, E, given by

$$E = nh\nu \text{ where } n = 1, 2, 3, \cdots \tag{1.20}$$

where $h = 6.626 \times 10^{-34}$ J.s. is Planck's constant and ν is frequency of the electromagnetic radiation. This is what is now referred to as the *Planck's quantum hypothesis*. A single quanta of light of energy $E = h\nu$ is known as a *photon*.

(ii) An ingeneous interpolation between the high frequency radiation spectrum and the low frequency radiation spectrum to obtain a radiation spectrum over the whole frequency range in terms of an energy density $U(\nu, T)$ per unit frequency interval or the radiant power per unit frequency interval was done by Planck, and the functions are given in several forms, such as

$$\frac{dU(\nu, T)}{d\nu} = \frac{8\pi h\nu^3}{c^3} \frac{1}{e^{h\nu/k_B T} - 1} \tag{1.21}$$

as the *energy density* of a photon gas, or as

$$\frac{dW(\nu, T)}{d\nu} = \frac{2\pi h\nu^3}{c^2} \frac{1}{e^{h\nu/k_B T} - 1} \tag{1.22}$$

as the *radiant power* of a black body, and it can be noted that equations (1.21) and (1.22) differ by a factor of $c/4$.

Equations (1.21) and (1.22) are different forms of *Planck's radiation law*, where $k_B = 1.381 \times 10^{-23}$ J.K^{-1} is the Boltzmann constant and all other symbols are as defined earlier. The frequency dependence of the energy density and the radiant power are illustrated in Figure 1.8 for three different temperatures.

We have somewhat given a historical development of the ideas developed to understand the black body radiation. Following the development of quantum mechanics in the early part of the 20th Century, equations (1.21) or (1.22) can now be proved explicitly by invoking ideas of quantum theory whereby light is made of particles known as *photons* and that these particles behave as *bosons* satisfying Bose-Einstein statistics.

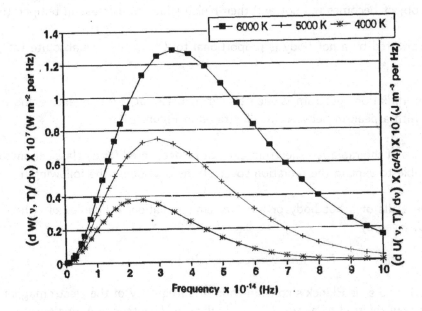

Figure 1.8: The frequency dependence of the energy density and radiant power at three different temperatures of a hot body (After Jain and Sharma).

All the observations of the radiation spectrum which could not be explained by classical physics can be explained by the Planck's radiation law. The peak in the radiation spectrum in terms of wavelength is found to satisfy

$$\lambda_m T = 2.9 \times 10^{-3} \text{mK} \tag{1.23}$$

which is known as *Wien's displacement law*, and explains observation (i) above. This means that if there are peaks at wavelengths λ_1, λ_2 and λ_3 at temperatures T_1, T_2 and T_3 respectively, then

$$\lambda_1 T_1 = \lambda_2 T_2 = \lambda_3 T_3 = \text{Constant} = 2.9 \times 10^{-3} \text{mK}$$

and that if $T_1 < T_2 < T_3$, then $\lambda_1 > \lambda_2 > \lambda_3$ (or $\nu_1 < \nu_2 < \nu_3$ in terms of frequencies).

The total energy density of a photon gas is the area under the radiation spectrum which is obtained by performing an integration, and the result is given by

$$U(T) = \sigma_g T^4 \tag{1.24}$$

and the total radiant power is obtained as

$$P = A\epsilon\sigma T^4 \tag{1.25}$$

where it can be noted that

$$\sigma = \frac{c}{4}\sigma_g$$

Equations (1.24) and (1.25) are alternative forms of *Stefan-Boltzmann law* , where A is the surface area of the emitting body, ϵ is the emissivity $(0 < \epsilon \leq 1)$, $\sigma = 5.67 \times 10^{-8}$ Wm^{-2}K^{-4} is the Stephan-Boltzmann constant, and T is the absolute temperature. This explains experimental observation (ii) above. The shape of the radiation spectrum is explained by equation (1.21) or (1.22) which invokes the particle nature of light. A hot body with $\epsilon = 1$ is a perfect emitter known as a black body.

It is worth noting the following limits of physical interest. In the low frequency range, the radiation spectrum reduces to:

$$\begin{aligned}
U(\nu, T) &= \frac{8\pi\nu^2}{c^3}\frac{h\nu}{\left[1 + \frac{h\nu}{k_B T} + \ldots - 1\right]} \\
&\approx \frac{8\pi\nu^2}{c^3}\frac{h\nu}{\left[1 + \frac{h\nu}{k_B T} - 1\right]} \\
&\approx \frac{8\pi\nu^2}{c^3}k_B T \tag{1.26}
\end{aligned}$$

and equation (1.26) is known as *Rayleigh-Jeans law* , first proposed in 1900. Rayleigh-Jeans law agrees with experimental observations fairly well at low frequencies but does not agree with experiment at high frequencies.

In the high frequency range, the radiation spectrum reduces to:

$$\begin{aligned}
U(\nu, T) &\approx \frac{8\pi h\nu^3}{c^3}e^{-h\nu/k_B T} \\
&\approx \nu^3 \alpha e^{-\beta\nu/T} \tag{1.27}
\end{aligned}$$

where α and β are constants, and equation (1.27) is known as *Wien's law*, first proposed in 1894. Wien's law agrees with experimental observations fairly well at high frequencies but does not agree with experiment at low frequencies.

It is interesting that Rayleigh-Jeans law and Wien's law were known long before quantum theory, but as discussed above these two laws were valid as approximations in the high and low frequency limits respectively. With the advent of the quantum theory, Planck's radiation law, given as equation (1.21) or (1.22), incorporated both these laws.

The photoelectric effect

In the *photoelectric effect*, when light is incident onto a metal, electrons are emitted under certain conditions. A typical experimental set up to observe the photoelectric effect is illustrated in Figure 1.9.

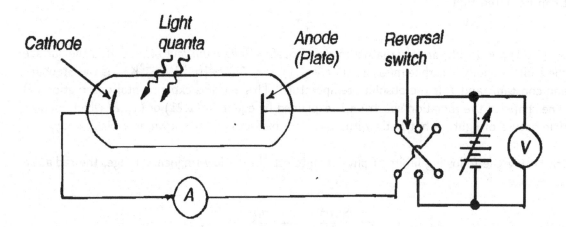

Figure 1.9: Experimental set up for observing the photoelectric effect. When light of frequency ν strikes the cathode plate, electrons are emitted and attracted to the anode plate, thus constituting an electric current.

Experimental observations show that:

(i) If the anode plate is made *positive* with respect to the cathode plate, electrons are accelerated towards the anode plate, and a photoelectric current is generated. This photoelectric current saturates to a constant value, I, depending on the intensity of the light. This implies that the intensity of incident light only determines the number of emitted photoelectrons, not their energy. Typical results are illustrated in Figure 1.10.

(ii) If the anode plate is made *negative* with respect to the cathode plate, electrons are repelled from the anode plate, and hence the current decreases rapidly until it becomes zero at some stopping voltage, V_s, independent of the intensity. Thus the maximum energy of a photoelectron is given by

$$eV_s = \frac{1}{2}mv_{max}^2 \tag{1.28}$$

(iii) For each metal, there exists a threshold frequency, ν_0, below which no electrons can be emitted no matter what the intensity of the incident radiation is. Typical results are illustrated in Figure 1.11. This implies that the energy of the photoelectrons emitted depends on the frequency of the incident radiation and the material constituting the cathode plate.

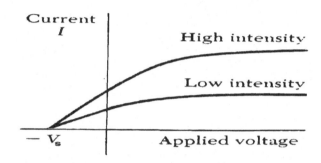

Figure 1.10: A graph of the photoelectric current, I against the applied voltage, V, for two intensities of light. The current increases with intensity but reaches a saturation value for positive values of V. The current vanishes for voltages equal to or less than $-V_s$.

The above observations of the photoelectric effect can not be accounted for by classical physics. According to classical physics, an increase in the intensity of the incident light should cause more electrons to be emitted, and a decrease in intensity should cause less electrons to be emitted, in agreement with observation (i) above. However, the relation between frequency of the incident radiation and the energy of the emitted electrons can not be explained by classical physics. Also, the existence of the threshold frequency as in observation (iii) above can not be accounted for by classical physics.

The failure of classical physics led to Einstein's theory of the photoelectric effect developed in 1905, and this is summarised below.

(i) Electromagnetic radiation consists of *quanta* which are discrete particles of light known as *photons*, with energy E, given by

$$E = nh\nu \text{ where } n = 1, 2, 3, \cdots$$

as discussed earlier, in equation (1.20).

(ii) When electromagnetic radiation interacts with the metal, it behaves as consisting of photons which are *particle like*, with each photon delivering energy to an individual electron, not to the atom or the metal as a whole.

(iii) The threshold frequency occurs because a certain amount of minimum energy must be supplied to the electron in order to free it from the metal.

(iv) Conservation of energy is satisfied. An incident photon of energy $h\nu$ is expended to remove an electron from a metal by doing some work (work function ϕ) and give the electron some kinetic energy $\frac{1}{2}mv^2$, and hence

$$h\nu \;=\; \phi + \frac{1}{2}mv^2 \tag{1.29}$$

Using equation(1.28) gives

$$h\nu \;=\; \phi + eV_s$$

$$\text{or}$$

$$V_s \;=\; \frac{h}{e}\nu - \frac{\phi}{e} \tag{1.30}$$

which can be plotted as V_s against ν, and will give a straight line of slope $= \frac{h}{e}$ and intercept $-\frac{\phi}{e}$. This is illustrated in Figure 1.11.

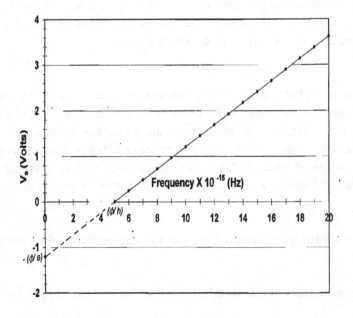

Figure 1.11: A graph of stopping voltage, V_s against frequency, ν. The straight line has a slope $= \frac{h}{e}$ and an intercept $-\frac{\phi}{e}$. Photons with frequency less that $\nu_0 = \frac{\phi}{h}$ do not have sufficient energy to eject electrons from plate.

X-ray production

In *X-ray production*, it is observed that when electrons bombard a metal target under certain conditions, X-rays (or *bremsstrahlung*, a word from German: *bremse* - brake and *strahlung* - radiation) are produced. A typical experimental set up for X-ray production is illustrated in Figure 1.12, where a metal coil C is heated and emits electrons of charge e by thermionic emission. These electrons strike a metal target A, and X-rays are emitted. The target A is held at a high positive potential in the range of kV to impart high energy to the incident electrons.

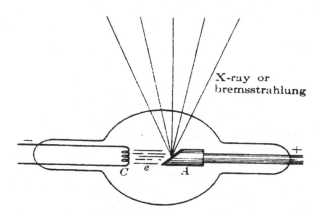

Figure 1.12: Experimental set up for X-ray (or bremsstrahlung) production.

Experimental observations show that:

(i) For each given electron energy, there is a minimum wavelength of the X-rays emitted.

(ii) The minimum wavelength of the X-rays emitted is independent of the type of metal used as a target material.

The above observations for X-ray production can not be explained by classical physics. According to classical physics there is no reason why there should be a minimum wavelength for the emitted X-rays.

The failure of classical physics to explain the minimum wavelength led to the *Duane-Hunt law*, developed in 1915, and this is summarised as follows. According to quantum theory, the minimum wavelength λ_{min} corresponds to the conversion of the total kinetic energy of a single electron into a single photon of the emitted X-rays, related by

$$eV = h\nu_{max} = \frac{hc}{\lambda_{min}} \qquad (1.31)$$

which can be expressed as

$$\lambda_{min} = \frac{hc}{eV} \qquad (1.32)$$

which is known as Duane-Hunt law, where h is Planck's constant, c is the speed of light, e is electronic charge and V is the potential difference across the X-ray tube.

The Compton effect

The *Compton effect* is observed when X-rays or γ-rays are scattered by electrons in matter. A schematic diagram for the Compton effect is illustrated in Figure 1.13.

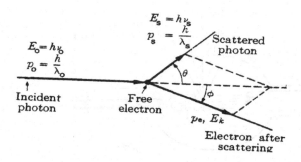

Figure 1.13: A schematic diagram illustrating the Compton effect.

Experimental observations show that:

(i) The scattered beam is observed to have two wavelengths, which are the incident wavelength λ_0 and the scattered wavelength λ_s, shifted from λ_0.

(ii) The shift, $\Delta\lambda = \lambda_s - \lambda_0$, is referred to as the *Compton shift*, and is independent of the target material, and only depends on the scattering angle θ.

The above observations for the Compton effect can be accounted for when it is invoked that the scattering can be explained by assuming that x-ray particles (photons) collide with electrons as particles. The Compton shift was first explained by Compton in 1922, and using conservation of momentum and conservation of energy, it is given by

$$\Delta\lambda = \lambda_s - \lambda_0 = \frac{h}{m_0 c}(1 - \cos\theta) \qquad (1.33)$$

where $h/(m_0 c)$ is known as the Compton wavelength.

1.4.4 Wave-Particle duality of light

In sections 1.4.1 and 1.4.2, the discussion showed that light is made up of waves. In section 1.4.3, the discussion concluded that light is made up of particles known as photons. These observations raise the question: what is light? Is light made up waves? Is light made up of particles? To reconcile these observations, light is actually *both*: at times it acts as if it is made up of waves while for some other physical phenomena it acts as if it is made up of particles, and hence the concept of *wave-particle duality*. What this means is that waves show particle-like properties and particles show wave-like properties. The concepts of De Broglie's hypothesis and Planck's energy quantisation are important in understanding the wave-particle duality.

The *De Broglies's hypothesis*, first postulated in 1924, states that momentum p (a particle property) and wavelength λ (a wave property) are related by

$$p = \frac{h}{\lambda} \tag{1.34}$$

Alternatively, this equation can be expressed as

$$p = \hbar k \tag{1.35}$$

where $\hbar = h/2\pi = 1.055 \times 10^{-34}$ J.s. and $k = 2\pi/\lambda$ is the wavenumber.

Planck's energy quantisation states that energy E and frequency ν are related by (recall equation (1.20))

$$E = h\nu \tag{1.36}$$

Alternatively, this equation can be expressed as

$$E = \hbar\omega \tag{1.37}$$

where $\omega = 2\pi\nu$ is the angular frequency. There is yet another alternative form, after using equations (1.2) and (1.34) in equation (1.36): it is given by

$$E = pc \tag{1.38}$$

1.4.5 The Einstein's mass-energy relation

One of the most important relations in physics is due to Einstein, whereby in his theory of relativity, mass m and Energy E are related by

$$E = mc^2 \tag{1.39}$$

where c is the velocity of light. This is the famous *Einstein mass-energy relation* which states that mass can be converted to energy, and energy can be converted to mass, and that in any physical process *mass plus energy* are conserved.

1.5   Chapter 1 Summary

Symbols used in the Chapter summary have the usual meaning as defined in the Chapter, and it holds for all the Chapter Summaries that follow each Chapter.

- Light travels along a straight line, known as the rectilinear propagation of light.

- Path of propagation of light is represented by a straight line with an arrow ahead pointing towards the direction of propagation, known as a ray of light.

- The rectilinear propagation of light is evidenced by image formation and shadow formation. Examples discussed are pin-hole camera, and the solar and lunar eclipse.

- Two classical methods, Fizeau's method and Michelsons's method are described for the measurement of the speed of light.

- Speed of light, c, in vacuum (free space) is the highest known speed in nature, given as $c = 2.997925 \times 10^8$ m.s$^{-1} \approx 3.0 \times 10^8$ m.s^{-1}.

- Light has a dual nature, wave nature and the particle nature, that is known as wave-particle duality.

- Electromagnetic waves satisfy the wave equation $\nabla^2 \vec{E} = \frac{1}{c^2} \frac{\partial^2 \vec{E}}{\partial t^2}$ and a similar equation for $\vec{B}$

- As per the wave nature, light is Electromagnetic (EM) waves which are transverse waves of wavelength λ and frequency ν which are related as $c = \lambda \nu$.

- Electromagnetic spectrum comprises radio waves, microwaves, infrared, visible, ultraviolet, x-rays, and gamma rays of successively increasing frequency (decreasing wavelength). Visible component is only a very small fraction of the entire EM spectrum.

- Wave nature of light is demonstrated by reflection, refraction, dispersion, interference, diffraction and polarization of light discussed in later chapters of the book.

- *Wien's displacement law* states that the peak in the radiation spectrum is found to satisfy $\lambda_m T = 2.9 \times 10^{-3}$mK

- *Stefan-Boltzmann law* states that the total power emitted by a body is given by $P = A\epsilon\sigma T^4$

- The *De Broglies's hypothesis* states that momentum p (a particle property) and wavelength λ (a wave property) are related by $p = \frac{h}{\lambda}$

- *Planck's energy quantisation* states that energy E and frequency ν are related by $E = h\nu$

- Light comprises photons of energy $E = nh\nu$.

- Properties of light such as radiation spectra of hot bodies, photoelectric effect, x-ray production and Compton effect can only be explained in terms of the photon nature of light which otherwise are inexplicable from the wave nature of light.

- Lastly, there is the Einsteins mass-energy relationship: $E = mc^2$.

1.6 Exercises

1.1.(a) How long does light from the sun take to reach the earth, given that light travels at a speed of 3×10^8 m/s and the earth-sun distance is 1.496×10^{11} m.
(b) Calculate the distance travelled by light in one year(known as a *Light Year*) .

1.2. Assuming that the Michelson's experiment is repeated at a location where the distance between the rotating mirror and the reflecting mirrors is 38 km, calculate the number of revolutions per minute of the eight sided mirror.

1.3. Radio stations broadcast at the following frequencies:
FM: 90.7 MHz; 93.36 MHz
MW: 621.0 KHz, 972.0 KHz
Convert these frequencies into wavelengths.

1.4. (a) When the Earth absorbs solar energy, it becomes heated, and subsequently re-radiates the energy into the atmosphere. At what wavelength is the radiation most intense if the Earth's surface temperature is 27 °C ?
(b) Calculate the temperature of a thermal source whose radiation has a maximum intensity when the wavelength is 1.8 μm.
(c) At what wavelength is the light emitted by the Sun most intense if the surface temperature of the Sun is approximately 5000 K?

1.5. The tungsten filament of an incandescent light bulb is a wire of diameter 0.08 mm and length 5.0 cm, and it is at a temperature of 3200 K. What is the power radiated by the filament assuming that it acts like a blackbody?

1.6. Calculate the energy (a) in J, (b) in eV, of a photon of blue light of wavelength 450 nm.

1.7.(a) Under certain conditions the retina of the human eye can detect as few as five photons of bluish green light of wavelength 5×10^{-7} m.
(i) Calculate the amount of energy received by the retina in Joules and in eV.
(ii) Suppose that the five photons are absorbed each second, what is the rate of energy transfer in watts?
(b) What is the mass equivalent of a single photon of wavelength 5×10^{-7} m?

1.8. Calculate the frequency, wavelength, momentum and mass equivalent of a photon with an energy of 2eV propagating in vacuum.

1.9. Electromagnetic radiation from interstellar clouds of hydrogen is detected at a 21 cm wavelength.
(a) calculate their frequency and photon energy.
(b) In which class of the electromagnetic spectrum is the detected radiation?

1.10. A proton, an electron and a photon have a wavelength of $1\overset{\circ}{A}$ each. If they all leave a given point at $t = 0$, what are the arrival times of each at a point 10m away?

1.11. Complete the following table

	Energy(eV)	λ (m)
	1	?
proton	?	1×10^{-10}
	1000	?
	1	?
electron	?	1×10^{-10}
	1000	?
	1	?
photon	?	1×10^{-10}
	1000	?

1.12. A mercury arc lamp is used as a source of UV radiation to study the photoelectric effect on lithium. By using various filters, the following wavelengths were used and the corresponding voltages, V_s, necessary to stop the photoelectrons completely were determined.

$\lambda(\overset{\circ}{A})$	2536	3132	3663	4358
$V_s(V)$	2.4	1.5	0.9	0.35

(a) Plot a graph of V_s versus frequency, ν.
(b) Hence, find
(i) the work function of lithium.
(ii)the value of h/e graphically, and compare it with the value computed from the values of the constants.

1.13. (a) Calculate the magnitude of the Compton wavelength, $h/(m_0 c)$.
(b) X-rays of wavelength 0.2 nm are scattered from a specimen of carbon in a direction $60°$ to the incident beam. Calculate the Compton shift, $\Delta\lambda$ and rhe wavelength of the scattered radiation.

1.14. (a) Sun radiates like a blackbody at 5000 K. Calculate the total power radiated by the sun, and the power received per unit area on the earth's surface. What is the total power received on total surface of the earth?
(b) Calculate the rate at which sun is losing mass.

1.15. Surface of a healthy human body is maintained at 36 ^oC. Model the body as a cylindrical object of height 1.6 m and diameter 0.4 m, with emissivity of 0.5. How much food in calories a person must eat every day just to compensate for the radiative loss of energy from the body in one day (1 cal = 4.2 J). Is your answer within the average daily food intake of a person? Discuss any discrepancies.

1.16. The maximum kinetic energy of an electron emitted from a metallic surface when radiation of 15×10^{14} Hz is used is 1 eV. Calculate the threshold frequency for the emission of photoelectrons from the metal.

1.17. Radiation of frequency 1.0×10^{15} Hz liberate electrons with a stopping potential of 1 V. Calculate the frequency of radiation that shall liberate electrons with a maximum kinetic energy of 6 eV from the same metal.

1.18. Capella a giant star, 4.3×10^{17} m away from earth radiates like a blackbody at 5200 K. The power on earth received from Capella is measured to be $1.2 \times 10^{-8} W m^2$. Assuming no loss of energy due to earth's atmosphere, calculate the radius of Capella.

1.19. X-rays of 40 KeV undergo Compton scattering. What is the minimum energy of the scattered X-rays, and at what angle of scattering it is detected?

1.20. Verify that the expressions for the electric field vector of the electromagnetic wave given by equation (1.19) satisfy the one dimensional wave equation (1.18).

Chapter 2

Laws of Geometrical Optics

When light traveling in a medium is incident on a boundary, depending on the nature of the boundary and the media it separates, the light may undergo reflection, refraction, scattering, and dispersion. Geometrical optics deals with the laws that govern these processes, and the application of these processes gives a number of optical instruments such as microscopes, telescopes, cameras, projectors etc. The laws of geometrical optics do not make any assumptions about the nature of light. Light propagating through a medium along a straight line travels with the speed of light in the medium, and is represented by a straight line with an arrow showing the direction of propagation. Such a graphical representation of light is known as the *ray of light*. A group of rays is termed as the *beam of light*, which may be a parallel, or a convergent or a divergent beam of light (Figure 2.1).

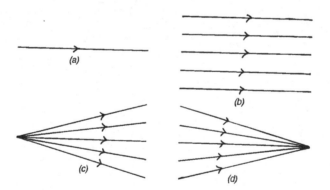

Figure 2.1: (a) A ray of light, (b) a parallel, (c) a divergent, and (d) a convergent beam of light.

The parallel beam of light appears to converge to or diverge from a point at infinity, (a very large distance compared to the focal length in the case of focusing optical components), for example rays from sun can be regarded as a parallel beam of light. Another typical and practical example commonly experienced are the parallel tracks of a railway line which appear to converge at a distance and also appear to diverge from a distant point. In geometrical optics we heavily rely on *ray diagrams* to locate, and characterize images formed when light is reflected or refracted by an optical component such as a mirror, or a lens, or when light is deviated or dispersed by a prism. Mathematical expressions to analytically locate and characterize images, and to calculate other optical parameters such as the angle of deviation of light, refractive index etc. are also derived using the ray diagrams combined with geometrical considerations. This gives the name *Geometrical Optics* to this branch of optics. The laws of optics discussed in this chapter can also be derived analytically using the wave nature of light, for example, by the Huygen's wave theory or the electromagnetic wave theory.

2.1 Reflection of Light

When light traveling in a medium impinges on a polished surface such as a mirror, it is bounced back in the same medium. The phenomenon known as *reflection of light* is governed by *two laws of reflection*. In order to define and discuss the laws of reflection, one first needs to understand the following terminology. *Reflecting surface (M)* is a polished, mirror surface which could be plane, curved or any shape from which an *incident ray (I)* traveling in a medium on incidence on the surface at the *point of incidence (P)* is bounced back in the same medium. The ray traveling back in the same medium after reflection from the surface is called the *reflected ray (R)*. The angle of the incident ray measured from the *normal (N)* at the point of incidence is the *angle of incidence (i)*, and the angle measured from the normal to the reflected ray is called the *angle of reflection (r)*. Angles *i* and *r* can take values from $0^o to \leq 90^o$. In a ray diagram, the cross section of a mirror with its reflecting surface perpendicular to the plane of paper is represented by a smooth line with shading on the non-reflecting side of the line. (Figure 2.2).

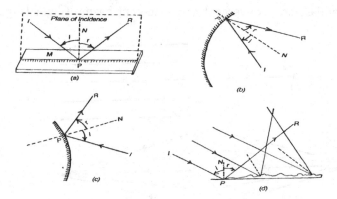

Figure 2.2: Reflection of light from (a) a plane mirror, (b) a concave mirror, (c) a convex mirror and (d) a rough reflecting surface.

Laws of reflection

First Law: Angle of incidence is equal to the angle of reflection, *i.e.* $\angle i = \angle r$.

Second Law: The incident ray, the normal to the reflecting surface at the point of incidence and the reflected ray lie in the same plane.

From geometrical considerations we know that any two lines meeting at a point lie within the same plane, for example the incident ray and the normal meeting at the point of incidence in this case, lie within the *plane of incidence* (Figure 2.2). The second law simply states that the reflected ray lies within the plane of incidence.

Laws of reflection apply to all reflecting surfaces irrespective of their shapes including uneven (rough) surfaces, provided the normal to the reflecting surface is at the point of incidence, and the angles are measured from the normal. Four cases of reflection from different surfaces are shown in Figure 2.2. However, in the case of an uneven surface, normals to the surface at various points of incidence are in random directions. Therefore, the reflected rays from an incident beam of light are *scattered* in all possible directions which can not be focused to provide an image of the source of the incident beam (Figure 2.2 d). Furthermore, from an opaque coloured object, there is a preferential reflection of light of the same colour as the colour of the object, and all other colours in the beam of an incident white light are absorbed. This explains the *colour of objects* as seen from the light reflected from the object. At night on wet roads, one experiences more glare from the on coming vehicles. Why?

2.1.1 Location and Characteristics of Image from a Plane Mirror

In order to locate the image of a point one must use at least two incident rays originating from the point. The two rays after reflection from a mirror (*i.e.* the reflected rays) either appears to diverge from a virtual point behind the mirror as in the case of a plane mirror, or they converge to a real point in space in front of the mirror as we shall see for some cases of a concave mirror discussed in Chapter 3. The point from which the reflected rays appears to diverge, or converges to is the *image* of the point from which the incident rays of light originated. If the reflected rays appear to diverge from a point behind the mirror the image is a *virtual image* which can not be taken on a photographic film or projected on a screen. On the other hand if the reflected rays converge to a real point, the image is a *real image* which can be taken on a photographic film or can be projected on a screen placed at the point of convergence. Real images are formed by certain spherical mirrors and lenses and their combinations. Photographic camera, slide or film projectors, and photocopying machines are some examples of real image formation which are discussed in later Chapters. In this section we shall concern ourselves only with the formation, location and characterization of an image formed from a plane mirror.

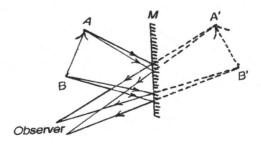

Figure 2.3: Image formation from a plane mirror.

Figure 2.3 shows the ray diagram to locate the image of an object *AB* formed from a plane mirror. We have used two incident rays from each of the two end points *A* and *B* of the object which after reflection appear to diverge from points *A'* and *B'* respectively. Hence *A'B'* is the virtual image of object *AB*.

Characteristics of the image:

From the ray diagram (Figure 2.3) and from our common experience with plane mirrors we note the following five characteristics of the image. Proof of some of these characteristics from the ray diagram are left as exercises

1 The image is virtual. The reflected rays appear to diverge from a point behind the mirror, which is the virtual image of the point from which the corresponding incident rays originate.

2 The image is as far behind the mirror as the object is in the front (Exercise 2.2).

3 The image is the same size as the object (Exercise 2.3). Thus the magnitude of *magnification* which is the ratio of the lateral size of the image to the lateral size of the object is one for a plane mirror.

4 The image is upright. This is a common experience from our reflection seen in a bathroom mirror.

5 The image is laterally inverted. This is also a common experience whereby our left is the right of the image and vice versa. On the front of ambulances, the word AMBULANCE is written left to right, with each of the letters reversed, Why?

2.1.2   Examples and Applications of Reflection from Plane Mirrors

Minimum Length of the Mirror to see full Image of oneself

The minimum length of a mirror required for a person to see his (her) complete, head to toe image is equal to half the height of the person, and the mirror should be mounted with its lower edge at a height from the ground which is equal to half the distance between toes and eyes of the person. This can be proved from the ray diagram shown in Figure 2.4, where *H*, *T* and *E* represent the head, toe and eye of the person, whose images in the mirror $M_1 M_2$ are *H'T'* and *E'* respectively as far behind the mirror as the person is in the front. For the sake of simplicity (clarity) we have used only one ray each from *H*, *T* and *E* reaching the eye after reflection from the mirror. Clearly, from the diagram $M_1 M_2$ is the minimum length of the mirror required to see the complete image, and it should be mounted with its lower edge at a height BM_2 from the ground.

Let *d* be the distance of the person from the mirror, and let $HT = h$ be the height of the person. Now in similar triangles $E\, M_1 M_2$ and $EH'T'$ taking the ratio of the sides and the heights of the triangles:

$$\frac{M_1 M_2}{H'T'} = \frac{EA}{EE'} = \frac{d}{2d}$$

$$\text{or} \quad M_1 M_2 \;=\; \frac{1}{2} H'T' = \frac{1}{2} HT = \frac{1}{2} h \qquad (2.1)$$

Similarly from similar triangles ETT' and $M_2 BT'$ one can show that

$$M_2 B = \frac{1}{2} ET$$

Note that these results do not depend on the distance of the observer from the mirror.

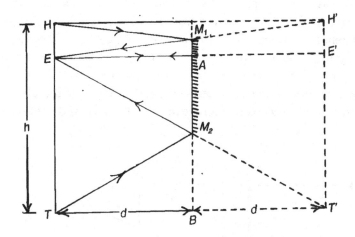

Figure 2.4: Full image of a person in a plane mirror as seen by himself (herself).

Reflection from a Rotating Mirror

Keeping the direction of the incident ray fixed, when a plane mirror is rotated by an angle θ about an axis in the plane of the mirror and perpendicular to the plane of incidence, the reflected ray is rotated by twice the angle, *i.e.* 2θ. The ray diagram is shown in Figure 2.5, and the simple proof is left as an exercise for students (Exercise 2.6).

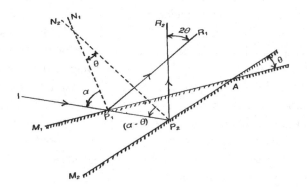

Figure 2.5: Reflection from a rotating plane mirror.

Reflection from two mirrors held at an Angle

When two plane mirrors are placed at an angle, with their reflecting surfaces facing each other, rays from an object placed between the mirrors undergo multiple reflections. On every reflection, the reflected beam appear to diverge from a virtual image in the corresponding mirror. This gives rise to multiple images of the object. The number of images so formed depend on the angle between the mirrors. Three cases are of particular interest that are discussed below.

Case(i): Two mirrors inclined at 90°: Figure 2.6 shows the ray diagram and formation of images. A set or rays (shown in the figure) after first reflection from mirror M_1 produces image I_1. The same set of rays after a second reflection from mirror M_2 produce image I_{12}. Likewise, another set of rays (not shown in the diagram) after first reflection from mirror M_2 produce image I_2, and after second reflection from mirror M_1 shall produce image I_{21}. The images I_{12} and I_{21} formed after second reflections, overlap each other, and one sees only three images of the object. As an exercise students are assigned to draw the second set of rays to show the formation of images I_2 and I_{21}.

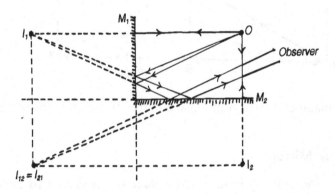

Figure 2.6: Two plane mirrors held at 90° produce three images of an object.

Case(ii): Two mirrors inclined at 60°: Figure 2.7 shows the ray diagram and formation of images. The set of rays (shown in the figure) which is first incident on mirror M_1 under goes three reflections, first from M_1, second from M_2, and the third one again from M_1. This produces three images, I_1, I_{12}, and I_{121}. Likewise, if we considered another pair of rays that under goes three reflections from M_2, M_1, and M_2 shall produce images I_2, I_{21}, and I_{212} respectively. Students should draw a ray diagram to show the formation of images I_2, I_{21} and I_{212}. The last two images from both the mirrors, *i.e.* I_{121} and I_{212} overlap, producing a total of 5 images. A toy, known as a *kaleidoscope* is based on this principle. Two or tree strips of plane glass or mirror are held at 60° in a tube with a few coloured glass pieces contained at one end of the tube which is closed with ground glass. On viewing through the other end, one sees colourful geometrical patterns formed from the multiple reflections of the coloured pieces by the glass strips.

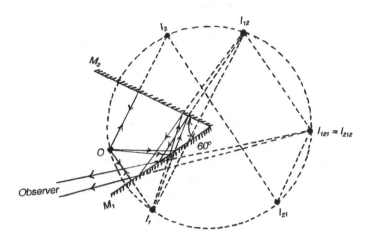

Figure 2.7: Images formed by two plane mirrors held at 60°.

Generally, if two mirrors are held at an angle θ^o, then the number of images *(n)* produced by the mirrors is given by:

$$n = \frac{360}{\theta} - 1 \qquad (2.2)$$

Case(iii): Two parallel mirrors facing each other, $(\theta = 0^o)$: The rays under go multiple reflections (theoretically infinite in number)successively from both the mirrors, producing an infinite number of images in each mirror (Figure 2.8). The figure shows the formation of only one set of images, I_1, I_{12}, I_{121}.... Draw a ray diagram to show the formation of the other set of images, I_2, I_{21}, I_{212}.... One also sees from eq. (2.2) that when $\theta = 0^o, n = \infty$. However, one sees only a finite but large number of images which is due to one or more of the reasons that follow: *(i)* As the images get farther and farther, their intensity diminishes due to loss in the intensity of light after each reflection if the mirror quality is not good, *(ii)* the visual limitation of the eye, and *(iii)* mirrors may not be perfectly plane or perfectly parallel.

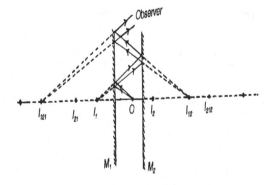

Figure 2.8: An infinite number of images are formed after multiple reflections from two plane parallel mirrors facing each other.

2.2   Refraction of Light

When light traveling in one medium enters another medium, on crossing the interface separating the two media it bends from its straight path. This effect is known as the *refraction of light,* and like reflection it is governed by *two laws of refraction.* The direction and magnitude of bending of the ray depends on the *optical densities* expressed in terms of refractive indices of the two media.

Refractive Index

Refractive index n of a refractive medium is a dimensionless physical property of the medium, defined as the ratio of the speed of light in vacuum *(c)* to the speed of light in the medium *(v),* i.e.,$n = (c/v)$. Since, $c > v$, n is always greater than 1. The speed of light in air is very close to c $(v_{air} \approx 2.989 \times 10^8 ms^{-1})$, so that the refractive index of air is very close to 1 $(n_{air} = 1.00029)$. Therefore, for all routine applications in geometrical optics we treat air like vacuum, and take its refractive index to be equal to 1.

For a given medium, the refractive index depends on the wavelength of light because the speed of light in a medium depends on wavelength. When light traveling in one medium enters another medium, its frequency remains the same (why?) but the wavelength changes. Let us say that the speed of light in the second medium is smaller than its speed in the first medium. Then the number of oscillations of light produced in one second (frequency of light) in the first medium, on entering the second medium shall 'crowd' together in a shorter distance, as a consequence of which the wavelength in the second medium is shortened. Since the speed, wavelength and frequency of light are related as: $v = \nu\lambda$, we can express the refractive index in terms of wavelength as:

$$ n = \frac{c}{v} = \frac{\nu\lambda_o}{\nu\lambda} = \frac{\lambda_o}{\lambda} \tag{2.3} $$

where λ_o and λ are the wavelength of light in vacuum and the medium respectively, and $\lambda_o > \lambda$ always.

Now, consider two media 1 and 2 with speeds of light v_1 and v_2, and refractive indices n_1 and n_2 respectively. If $v_1 < v_2$, then $n_1 > n_2$, and medium 1 is referred to as optically *denser* medium, and medium 2 is the *rarer* medium. Sometimes one uses the *relative refractive index* (n_{12}) of medium 1 with respect to medium 2 defined as:

$$ n_{12} = \frac{n_1}{n_2} = \frac{c}{v_1}\frac{v_2}{c} = \frac{v_2}{v_1} \quad or \quad n_1 v_1 = n_2 v_2 \tag{2.4} $$

Equation (2.4) can also be expressed in terms of the wavelength of light in two media as:

$$ n_{12} = \frac{n_1}{n_2} = \frac{v_2}{v_1} = \frac{\lambda_2}{\lambda_1} \quad or \quad n_1 \lambda_1 = n_2 \lambda_2 \tag{2.5} $$

Depending on the relative magnitudes of v_1 and v_2, n_{12}, may be greater or less than 1. As stated above, since $n_{air} \approx 1$, relative refractive index of any medium with respect to air may be taken to be the absolute refractive index of the medium for most of the routine applications, *i.e.* $n_{1,air} \approx n_1$.

For some materials the refractive index does not depend on the direction of propagation of light, whereas in some other materials it depends on the direction of propagation of light. These two categories of materials are known as *homogenous* and *heterogenous* medium respectively. Lastly, refractive index of certain materials also depends on the polarization of light, known as birefringence which is discussed in Chapter 8 on Polarization. For most part of Geometrical Optics, unless otherwise stated, we shall treat all refractive media to be homogenous, incident radiation to be unpolarized, and refractive index to be constant corresponding to some mean wavelength of radiation. Table 2.1 gives (absolute) refractive index of some common materials.

Table 2.1: Absolute refractive index of some common materials.

MATERIAL	n
Air	1.000293
Carbon dioxide	1.00045
Hydrogen gas	1.000132
Water	1.333
Ice	1.31
Carbon tetrachloride	1.461
Ethyl alcohol	1.361
Crown glass	1.52
Flint glass	1.66
Fused silica	1.458
Diamond	2.419

Refractive index: Cauchy and Sellmeier's equations

The wavelength (frequency) dependence of the refractive index given by equation (2.5) is rather a very simplistic picture. There are several formulations that give the quantitative dependence of the refractive index on wavelength. Here we shall present just two commonly used formulations due to Cauchy and Sellmeier.

Cauchy in 1836 gave two different relations for the refractive index $n(\lambda)$ in terms of the wavelength λ, known as Cauchy's equations. The first form is the two-constant Cauchy equation:

$$n(\lambda) = A + \frac{B}{\lambda^2} \tag{2.6}$$

where A, and B are constants which are characteristic of the refractive material. The second form is the three-constant Cauchy equation:

$$n(\lambda) = A + \frac{B}{\lambda^2} + \frac{C}{\lambda^4} \tag{2.7}$$

where A, B and C are constants which are also characteristic of the material.

Another formulation of the wavelength dependence of the refractive index is what is referred to as Sellmeie'r's equation, where the refractive index n satisfies

$$n(\lambda) = 1 + \frac{A\lambda^2}{\lambda^2 - \lambda_0^2} \qquad (2.8)$$

$$= 1 + \frac{A}{1 - (\lambda_0^2/\lambda^2)} \qquad (2.9)$$

It can easily be shown that Cauchy's equation is an approximation of Sellmeier's equation. This is left as an exercise for the student.

Refractive index and Dielectric function

The refractive index ($n(\omega)$ or $n(\lambda)$) of a material is related to the dielectric constant ϵ of the material, which in practically a function of wavelength (frequency), and we can therefore refer to it as a dielectric function, i.e., $\epsilon(\omega)$ or $\epsilon(\lambda)$. The relationship is rather simple, given as:

$$n(\omega) = \sqrt{\epsilon(\omega)} \qquad (2.10)$$

The frequency dependence of the dielectric function for insulators or semiconductors is given by

$$\epsilon(\omega) = \epsilon(\infty) + \frac{S\omega_T^2}{\omega_T^2 - \omega^2} \qquad (2.11)$$

where $\epsilon(\infty)$ is the high frequency dielectric constant, S gives the strength of the resonance, ω_T is the transverse optical phonons frequency, characteristic of the material.

The Dielectric function for metals, $\epsilon(\omega)$ is given by:

$$\epsilon(\omega) = 1 - \frac{\omega_p^2}{\omega^2}$$

where ω_p is the plasma frequency.

Laws of refraction

In order to understand refraction, and the laws of refraction we introduce additional *terminology* that follows. The point at which the ray enters the second medium is the *point of refraction (P)*. It coincides with the point of incidence of the incident ray at the interface. The ray on entering the second medium is called the *refracted ray (R)*. Angle of the refracted ray measured from the *normal(N')* to the interface at the point of refraction is the *angle of refraction (r)*, and the plane containing the refracted ray and the normal is the *plane of refraction*. From geometrical considerations, normals N and N' to the interface in the two media are along the same straight line *NPN'* (Figure 2.9).

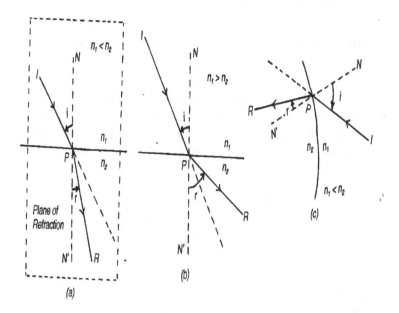

Figure 2.9: Refraction of light (a) from rarer to denser medium, and (b) from denser to rarer medium at a plane interface, and (c) from rarer to denser medium at a curved interface.

First Law - Snell's Law: When a ray of light traveling in a medium of refractive index n_1 enters another medium of refractive index n_2, the angles of incidence i and the angle of refraction r are related to the refractive indices n_1 and n_2 as:

$$n_1 \sin i = n_2 \sin r$$
$$\text{or } \frac{n_2}{n_1} = \frac{\sin i}{\sin r} \tag{2.12}$$

If medium 1 is air, and if we treat air to be a medium of refractive index 1, then the refractive index of the second medium may be taken as: $n_2 = (\sin i / \sin r) = n$. From equation (2.12), if $n_2 > n_1$, *i.e.* if medium 2 is a denser medium, then $\angle r < \angle i$. Thus a ray traveling from a rare medium in to a denser medium bends towards the normal (Figures 2.9 a and c). From the reversibility of the path of light, a ray traveling from a denser medium to rarer medium bends away from the normal, (*i.e.*, $\angle r > \angle i$) (Figure 2.9 b).

Second Law: The incident ray, the refracted ray, and the normal to the interface at the point of refraction lie in the same plane.

This simply implies that the plane of incidence in medium 1, and the plane of refraction in medium 2 are along the same common plane extending from medium 1 in to medium 2 across the boundary of the two media, and the incident ray, refracted ray and normal lie within this plane. The laws of refraction apply to any shape of the interface provided the normal to the interface at the point of refraction is considered appropriately (Figure 2.9 c).

2.2.1   Examples and Applications of Refraction at a plane interface

Refraction through a rectangular block

Consider a rectangular block of width t of a material of refractive index n, for example glass. A ray of light traveling in air is incident at an $\angle i$ at one of the sides of the block, and after passing through the block emerges in to air (refractive index n_a) from the opposite side of the block (Figure 2.10). *The emergent ray is parallel to the incident ray, and is displaced by a distance d with respect to the incident ray*, given by (Exercise 2.14):

$$d = t \sin i \left(1 - \frac{n_a \cos i}{\sqrt{n^2 - n_a^2 \sin^2 i}} \right)$$

(2.13)

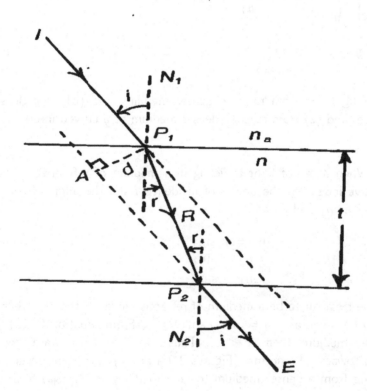

Figure 2.10: Refraction of light through a rectangular block of thickness t.

Apparent depth

It is a common experience that a coin at the bottom of a glass filled with water, and the bottom of a swimming pool appear to be raised, *i.e.*, the depth of these objects as seen from air appears to be less than the actual depth. Also, a spoon or a straw dipped in a glass of water appears to be bent upwards at the water air interface. These optical illusions happen because of the refraction of light from a denser medium (water) in to rarer medium (air), whereby a ray originating from a point in water on entering air bends away from the normal. The refracted ray then appears to come from an elevated point in water, and one sees a virtual image of the point at a lesser depth.

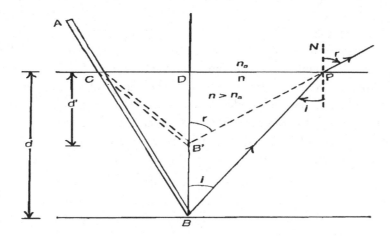

Figure 2.11: Apparent depth of a point inside a denser medium as seen from air.

Figure 2.11 shows the refraction of a ray from water into air. *B* is a point at the bottom of a pool of water, or the lower end of a straw *ACB* dipped in a glass of water at a depth *d*. A ray *BP* from *B* on entering air bends away from the normal *N*. On extending the refracted ray backwards it appear to diverge from a point *B'* which is higher up from the point *B* at the bottom. *B'* is the virtual image of point *B*, and its depth d' ($< d$) is the apparent depth one sees. Let *i* and *r* be the angles of incidence and refraction respectively, and n_a and *n* are the refractive indices of air and water respectively. From the figure:

$$\sin i = \frac{PD}{PB}, \quad \text{and} \quad \sin r = \frac{PD}{PB'} \tag{2.14}$$

From Snell's law, and equation (2.14):

$$\frac{n_a}{n} = \frac{\sin i}{\sin r} = \frac{PB'}{PB} \tag{2.15}$$

To view the depth one looks almost vertically down, in which case points *P* and *D* are very close together, and one can take $(PB')/(PB) \approx (DB')/(DB) = d'/d$. Thus the apparent depth is given as:

$$d' = d \, \frac{n_a}{n} \tag{2.16}$$

Since $n > n_a$, the apparent depth d' is smaller than the real depth *d*. This also makes the straw *ACB* appear bent at the water interface as *ACB'*.

Caution While entering a swimming pool, or allowing children to enter, one should be aware of the 'apparent depth' illusion which is smaller than the real depth.

Total internal reflection

For a ray of light traveling from a denser medium of refractive index n_d to a rarer medium of refractive index n_r, the angle of refraction r is larger than the angle of incidence i. Starting from a small value, if the angle of incidence is increased gradually, the angle of refraction, which shall always be larger than the angle of incidence, also increases till for certain value of the angle of incidence $(i = i_c)$ the angle of refraction becomes 90° (rays I_3 and R_3, Figure 2.12). In this case the ray on entering the rarer medium is refracted tangentially to the interface. If the angle of incidence is further increased, even by just a fraction of a degree, the ray does not enter the rarer medium, and it is reflected back from the interface in the denser medium as per the laws of reflection (rays I_4 and R_4, Figure 2.12). This is called the *total internal reflection*, as the reflection of the energy of the incident radiation is 100%, with none, whatsoever lost to the rare medium. The limiting (maximum) angle of incidence for which the ray is just able to enter the rarer medium, (for which the angle of refraction is 90°), is known as the *critical angle* (i_c). Thus for $i = i_c$, $r = 90^o$, and from Snell's law:

$$\frac{n_r}{n_d} = \frac{\sin i_c}{\sin 90^0} = \sin i_c$$

$$\text{or} \quad n_d = \frac{n_r}{\sin i_c} \tag{2.17}$$

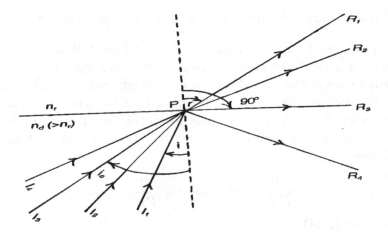

Figure 2.12: Total internal reflection and the critical angle.

Applications and consequences of total internal reflection

1. Optical fiber is a thin, flexible wire of glassy material, a bundle of which is used to direct light into inaccessible places such as into human body to examine and to photograph the internal organs for medical diagnostics. They have also found extensive applications in modern telecommunication. Some 10 billion digital bits per second, equivalent to tens of thousands of telephone calls, can be transmitted per second along an optical fiber, thus rendering both high speed and high capacity to optical-fiber communication compared to the conventional communication means. The transmission of light signals through the cable is based on the principle of total internal reflection. A typical fiber consists of high purity silica glass enclosed by a protective sheath. The outer layer, termed cladding has a smaller refractive index than that of the inner core layer, and the core is characterized by a critical angle i_c of total internal reflection. A light ray (signal) that enters the core of the fiber through the cross section is incident at the core-cladding interface at an angle $i \geq i_c$. The signal undergoes multiple internal reflections within the fiber till it emerges and received at the other end of the cable (Figure 2.13). Because of the *'total internal reflections'* there is no loss in the power of the signal. This is just a simple principle of how optical fibers function. A variety of optical fibers with varied specifications have been engineered for commercial applications. They include a fiber with a very narrow core (8 micron or less) known as the single mode fiber with a smaller difference in the refractive indices of the core and the cladding; graded-index multimode fiber for which the refractive index gradually decreases outwards from the axis of the core to the cladding; and a step-index multimode fiber with a much wider core of the order of 100 micron.

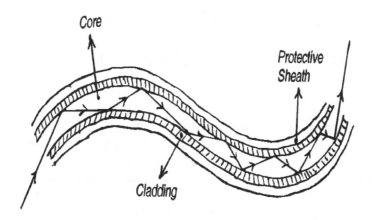

Figure 2.13: Transmission of light through an optical fiber.

2. Fish eye view of our world: A ray from air on entering water is bent towards the normal, and all the rays from air, from horizon to horizon, after refraction are confined within a cone of solid angle $2\,i_c$. Thus, our entire world above water lying within a hemisphere of solid angle $2\,\pi$ is imaged within the cone of vision of solid angle $2i_c$ in water. Thus fish is able to see horizon to horizon above water, although the vision is distorted (Figure 2.14). That is why it is always very difficult to catch a fish with bare hands, no matter how stealthily one sneaks on it. The success depends on the speed and the element of surprise. Furthermore, a smaller apparent depth of the fish as seen by the fisherman also contributes to low success rate in catching one with bare hands. This is from where the wide angle camera lenses also get their name *'fish eye lens'* which are used to take wide angle photograph extending almost horizon to horizon horizontally, and from the sky to the ground vertically. However, the image is distorted.

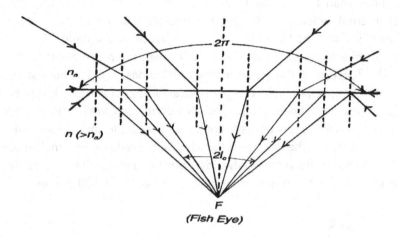

Figure 2.14: 'Fish eye-view' of our world.

3. Displacement and bending of light in optical instruments: In some optical instruments for specialized applications such as periscopes, telescopes, microscopes etc., one needs to displace the incoming beam, or bend it by 90° or 180°. This is achieved by using right-angle glass prisms in conjunction with the principle of total internal reflection. Some examples are shown in Figure 2.15.

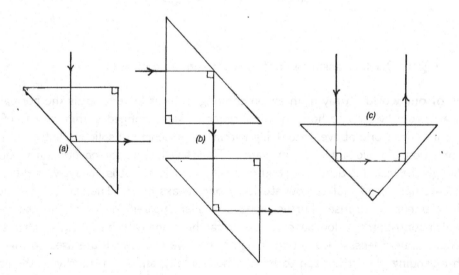

Figure 2.15: Bending and displacement of light using right angle prisms.

4. Mirages are the optical illusions seen in a hot desert, and at cold sea that result from refraction of light from a denser medium to rarer medium and subsequent total internal reflection. In a hot desert one sees the image of a distant tree (object) as if the tree was reflected in a pool of water, and at cold sea one sees the 'ghost' image of a distant ship below horizon hanging upside down in air (Figure 2.16).

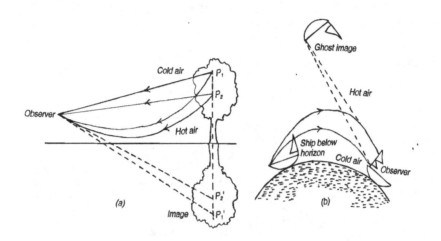

Figure 2.16: Mirage (a) in desert, and (b) at sea.

In the desert during the day, layers of air close to earth's surface are hotter and rarer as compared to the upper layers, and there is a gradual increase in the temperature and decrease in optical density of air from high above to below as one goes down towards the earth's surface. Therefore, a ray of light from top of a tree traveling towards an observer is successively bent away from the normal as it passes through hotter, and rarer layers of air. Finally, at some point along its path, where the angle of incident becomes larger than the critical angle, it undergoes total internal reflection. This is the lowest point in the path of the ray, and from here it travels towards the observer passing from a rarer layers of air to denser layers, bending towards normal at each step of the way. Thus the ray reaching the observer, on extending backwards appears to come from a virtual image of the tree causing an illusion as if there was water at the foot of the tree. Travelers, and animals are known to have died in deep deserts in search of water because of this illusion. On the other hand, at sea it is the opposite of the process of refraction in desert. Air close to the sea surface is cold and denser than the air above. Therefore, the image of a distant ship is seen hanging upside down in the air. Ships are known to have been saved from pirates because the ghost-image of the pirate ship was seen before the ship itself became visible above the horizon.

2.3 Deviation and Dispersion of Light

To study the deviation and dispersion of light one uses a prism. A *Prism* is a block of refractive material with triangular cross section, two sides of which act as the refracting surfaces. The angle between the two faces of the prism through which the light passes is known as the *angle of prism* A. The third face, opposite A is the *base of the prism* on which the prism rests. Depending on application, the triangular cross section in principle may have any geometry; cross section of the most commonly used prisms is either an equiangular triangle $(A = 60^o)$, or a right angle triangle, or an isosceles triangle of which the two symmetrical sides are the refracting surfaces. Also, depending on the application, a prism may be made of glassy materials, or alkali halide crystals, or may be liquid filled. Here we shall restrict only to the use of symmetrical, solid glass prisms, either with the equiangular, or the isosceles triangular cross section.

2.3.1 Deviation of light

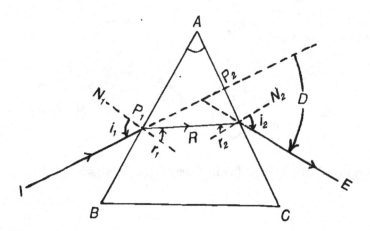

Figure 2.17: Deviation of light on refraction through a prism.

Consider a monochromatic (single wavelength) ray of light incident on the left refracting surface of a prism. On entering the prism, the angle of refraction is smaller than the angle of incidence so that the ray traveling through the prism is bent towards the base. At the second surface, traveling from a denser to a rarer medium, the emergent ray further bends towards the base. The angle by which the emergent ray is rotated with respect to the incident ray is known as the *angle of deviation (D)* (Figure 2.17). Besides the prism itself, the angle D also depends on the angle of incidence, and for a certain value of the angle of incidence, say i_o, the angle of deviation is minimum (D_{min}) (Figure 2.18). It is this angle of minimum deviation which is of prime interest in most applications, and it is related to the refractive index of the material of the prism (n) as:

$$\frac{n}{n_a} = \frac{\sin(\frac{A+D_{min}}{2})}{\sin(\frac{A}{2})}$$

(2.18)

where n_a is the refractive index of air in which the incident and emergent rays travel.

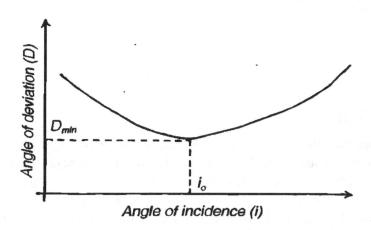

Figure 2.18: Angle of deviation as a function of the angle of incidence (schematic).

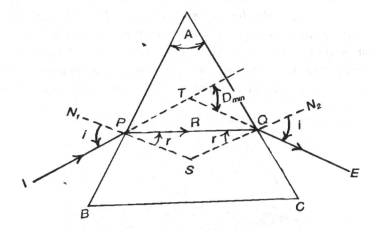

Figure 2.19: Refraction of light at minimum deviation through a symmetrical prism.

Proof: Figure 2.19 is the ray diagram showing the refraction of light through a symmetrical prism at minimum deviation. In a symmetrical prism (equiangular or isosceles cross section) the ray at minimum deviation passes through the prism symmetrically such that the ray going through the prism is parallel to the base of the prism, and the angle of emergence of the ray is equal to the angle of incidence i. The angles that the refracted ray through the prism makes with the normals at both the sides of the prism are also equal, say r. Let A be the angle of the prism.

In quadrilateral PAQS: $\angle PSQ = (180 - A)$
In triangle PQS: $\angle PSQ = (180 - 2\,r)$
Hence: $(180 - A) = (180 - 2r)$ or $r = (A/2)$

In triangle PTQ: $\angle TPQ = (i - r) = \angle TQP$,
and from the properties of a triangle:
$D_{min} = (\angle TPQ + \angle TQP) = 2(i - r) = (2i - A)$
Or $i = (A + D_{min})/2$
Finally, from Snell's Law:

$$\frac{n}{n_a} = \frac{\sin i}{\sin r} = \frac{\sin(\frac{A+D_{min}}{2})}{\sin(\frac{A}{2})} \qquad (2.19)$$

2.3.2   Dispersion of light

In Section 2.2 we discussed that the refractive index depends on the wavelength of light (equation 2.5 or 2.6 to 2.9). Hence from equation (2.19)the angle of minimum deviation from a prism shall also depend on the wavelength. When white light that consists of the entire range of wavelengths from violet to red is refracted through a prism, components with different wavelengths have different angles of deviation, and entire spectrum of white light is seen in the emergent beam. This is called the *dispersion of light*. Although dispersion also takes place when a beam of light passes through a rectangular glass block as in section 2.2.1, but in that case since all emergent rays are parallel to the incident rays, eye receives all the colours at the same angle that overlap, and are not resolved by the eye. Since the wavelength for violet light is shorter than the red wavelength, the refractive index for violet colour is larger than the refractive index for the red (equation 2.5). From equation (2.19) this implies that the angle of deviation for violet light is larger than that for the red light. Therefore, in the spectrum of white light produced by a prism, violet light is bent more towards the base, while the red light having bent least is away from the base, with all other colours in between (Figure 2.20). Angle θ between the two extreme red and violet rays of the spectrum is the angle of dispersion.

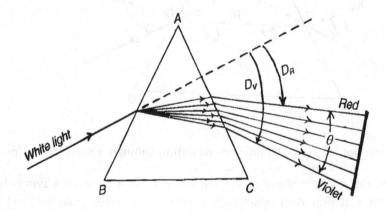

Figure 2.20: Dispersion of white light on refraction through a prism.

Applications of deviation and dispersion in Sciences and in Nature

In sciences deviation of light by prisms is used: *(i)* to determine unknown wavelengths of light if refractive index of the prism material as a function of wavelength is known, and*(ii)* by using light of known wavelengths, prisms of new materials are used to determine refractive index of the materials as a function of wavelength. In nature, rainbows are formed due to the dispersion of sun's light by small droplets of water in the atmosphere followed by internal reflection(s) before the coloured rainbow-beam emerges from the droplet (Figure 2.21).

For the primary rainbow , there is only one internal reflection of beam within the water droplet, so that the red colour is higher up and violet is the lowest in the rainbow. For the secondary rainbow, there are two internal reflections of the beam within the water droplet, and the colours in the secondary rainbow are reversed. Furthermore, the secondary rainbow due to two reflections is dimmer than the primary rainbow, and higher orders, if any, are so weak that they are not visible to the bare eye.

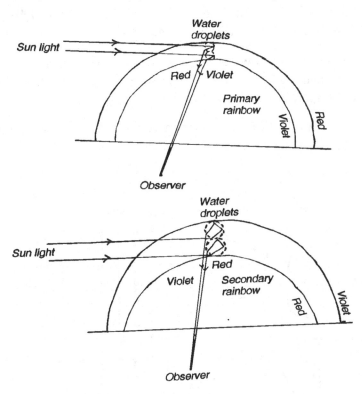

Figure 2.21: Formation of primary and secondary rainbows on dispersion of white light by water droplets.

2.4 Huygen's Wave Theory for propagation of light

Light emitted from a source travels in all directions in the medium. In case of a point source the rays from the source diverge from the source, and travel spherically symmetrically in all direction. As a consequence, light energy emitted from a source at $t = 0$ shall be found at the surface of a sphere of radius $r = t\,v$ at a later time t where v is the speed of light in the medium. In case of the radius of the sphere extending to ∞, a finite section of the sphere seen by an observer appears plane. In context of wave propagation, these spherical and plane surfaces, that are normal to the direction of propagation of waves represented by divergent and parallel beam of rays respectively are termed *wavefronts*. More precisely, wavefront is the locus of all points in space that oscillate with the same phase, *i.e.*, phase difference between any two points on a wavefront is zero. Because in a homogeneous medium light travels with the same speed in all directions, the successive positions of wavefronts in a medium are parallel to each other. Using the wave nature of light, and the concept of wavefronts Huygen gave a wave theory for the propagation of light in any medium. The salient features of the theory are:

- Waves propagate through a medium as wavefronts. All points on a wavefront are in phase.

- Every point on the wavefront is a source of secondary wavelets. These wavelets propagate in the forward direction with the speed of waves in the medium.

- A common envelope to the secondary wavelets gives the new position of the wavefront in the medium. Thus two successive positions of the wave at time interval Δt is given by two parallel wavefronts distance $v\,\Delta t$ apart in the forward direction.

Figure 2.22 shows the propagation of light from a point source (spherical wavefronts and divergent rays), and from a distance source (plane wavefronts and parallel rays).

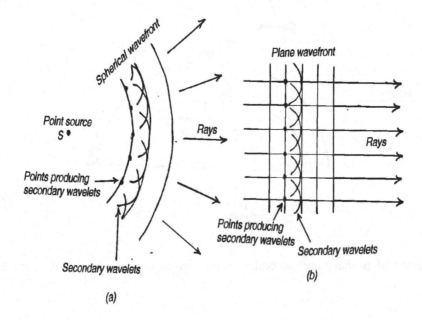

Figure 2.22: Rays, wavefronts and secondary wavelets: (a) divergent rays and spherical wavefronts, (b) parallel rays and plane wavefronts.

2.4.1   Application to refraction

Figure 2.23 shows the refraction of a parallel beam of light (only two extreme rays are shown) at the interface of two media of refractive indices $n_1 (= c/v_1)$ and $n_2 (= c/v_2)$ where v_1 and v_2 are the speeds of light in respective medium. The plane wave front $P_1\ A$ for the incident rays is perpendicular to the rays. At time $t = 0$ one end of the wavefront just touches the interface at the point of incidence P_1 of one of the rays. At this moment the secondary wavelet produced at P_1 shall propagate in the second medium with speed v_2, where as the secondary wavelets from A shall continue in the first medium with speed v_1. Δt time later when the wavelets from A would have reached point of incidence P_2 of the second ray, wavelets from P_1 would have reached point B in the second medium. Thus, $B\,P_2$ is the new, refracted position of the wavefront $\Delta\,t$ time later, and the refracted rays are perpendicular to the refracted wavefront in the second medium.

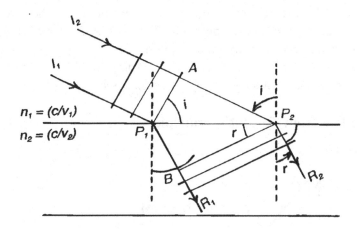

Figure 2.23: Refraction of a plane wavefront at the plane interface of two media: Snell's law from Huygen's wave theory.

From the speeds of light in the two media:

$A\,P_2 = v_1\,\Delta t.$
$P_1\,B = v_2\,\Delta t.$

If *i* and *r* are the angles of incidence and refraction respectively, then

In $\triangle P_1\,A\,P_2$: $\sin i = (A\,P_2/P_1\,P_2)$, and
In $\triangle P_1\,B\,P_2$: $\sin r = (P_1\,B/P_1\,P_2)$.

Hence: $(\sin i/\sin r) = (A\,P_2/P_1\,B) = (v_1/v_2) = (n_2/n_1)$,
or $n_1\,\sin i = n_2\,\sin r$

which is *Snell's law of refraction*, proved by Huygen's wave theory.

2.5   Fermat's Principle

Fermat's principle states that *light traveling from one point to another point, takes the path along which the time of travel is minimum.* The two points may lie within the same medium, or in two different media, or any number of media may lie in between. Reversibility of the path of light is a natural consequence of Fermat's principle. If a light traveling in one direction takes a particular path that has the minimum time of travel, then to travel in the opposite direction the ray must take the same path to take minimum time to travel. Therefore, in reflection, refraction, or deviation of light if one changes the reflected, refracted, or emergent ray to incident ray, then the incident ray changes to reflected, refracted or emergent ray respectively (Figures 2.2, 2.9, and 2.17). The laws of reflection and refraction can also be derived from Fermat's principle by applying to the time of travel of a ray between two points the basic principle of calculus according to which the first derivative of a quantity at minimum is zero. Here we shall explicitly prove Snell's law from Fermat's principle.

2.5.1   Proof of Snell's law from Fermat's principle

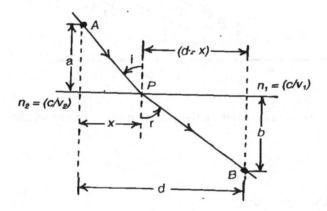

Figure 2.24: Fermat's principle and the Snell's law.

Figure 2.24 shows the refraction of a ray of light from a medium of refractive index $n_1(= c/v_1)$ in to a medium of refractive index $n_2(= c/v_2)$, where v_1 and v_2 are the speeds of light in these media. i and r are the angles of incidence and refraction respectively, and P is the point of incidence. Consider the propagation of light from point A in medium 1 to point B in medium 2. The time of travel is:

$$t = \frac{AP}{v_1} + \frac{PB}{v_2} = \frac{\sqrt{a^2 + x^2}}{c/n_1} + \frac{\sqrt{b^2 + (d-x)^2}}{c/n_2} \tag{2.20}$$

For fixed points A and B in two media, distances a, b and d are fixed, and the time of travel for the light depends on x only, *i.e.*, on point P where the ray crosses the interface of two media. Location of P also determines the angles of incidence and refraction. Therefore, we must minimize the time of travel with respect to x. Hence, applying the condition of minimum, *i.e.* $(dt/dx) = 0$, to equation (2.20)

$$\frac{dt}{dx} = 0 = \frac{n_1}{c} \frac{d}{dx} \sqrt{a^2 + x^2} + \frac{n_2}{c} \frac{d}{dx} \sqrt{b^2 + (d-x)^2}$$

$$= \frac{n_1 x}{c\sqrt{a^2 + x^2}} - \frac{n_2(d-x)}{c\sqrt{b^2 + (d-x)^2}}$$

$$\text{or} \quad \frac{n_1 x}{\sqrt{a^2 + x^2}} = \frac{n_2(d-x)}{\sqrt{b^2 + (d-x)^2}} \tag{2.21}$$

But from Figure 2.24:

$$\sin i = \frac{x}{\sqrt{a^2 + x^2}}, \quad \text{and} \quad \sin r = \frac{(d-x)}{\sqrt{b^2 + (d-x)^2}} \tag{2.22}$$

Therefore, from equations (2.21 and 2.22):

$$n_1 \sin i = n_2 \sin r \tag{2.23}$$

which is the Snell's law of refraction, proved by Fermat's principle.

In practical life one can apply Fermat's principle to a situation whereby a lifeguard on the beach must run part of the way along the beach and part swim in water to save a drowning swimmer at sea. Depending on the speeds of the lifeguard on ground and in water, there is an optimal point where the lifeguard must enter the water to take minimum time to reach the victim. However, note that when the emergency arises there would not be enough time to first calculate the optimal point of entry. Experience based judgment must be used by the lifeguard to act.

2.6 Chapter 2 Summary

- A beam of light is a group of rays of light that may be parallel, convergent or divergent. Rays from a distant source, for example the sun are taken to be parallel.

- Reflection is the bouncing back of a ray of light within the same medium when it strikes a polished (reflecting) surface. It is governed by two laws: (i) The angle of incidence is equal to the angle of reflection, and (ii) The incident ray, the normal to the reflecting surface at the point of incidence and the reflected ray lie in the same plane.

- Image formed from a plane mirror is: (i) Virtual, (ii) as far behind the mirror as the object is in front of the mirror, (iii) upright, (iv) same size as the object, and (v) laterally inverted.

- A virtual image is the one from which the rays of light appear to converge. Virtual image cannot be projected on a screen or taken on a photographic film.

- Keeping the incident ray fixed, if the mirror is rotated by an angle θ, the reflected ray is rotated by twice the angle, *i.e.* θ.

- When an object is placed between two mirrors held with their reflecting surfaces at angle θ, the number of images, n, formed is given by: $n = \frac{360}{\theta} - 1$. For $\theta = 90^o$, 60°, and parallel mirrors, $n = 3, 5$ and ∞ respectively.

- Refraction is bending of the ray of light from its straight path when it enters from one medium in to the other. It is governed by two laws. (i) Snells Law: $n_1 \sin i = n_2 \sin r$, and (ii) the incident ray, the normal to the interface at the point of refraction and the refracted ray lie within the same plane.

- A refractive medium is characterized by a refractive index, $n = (c/v)$. Note that $n > 1$ for all media.

- Refractive index of medium 1 relative to medium 2 is given by: $n_{12} = \frac{n_1}{n_2} = \frac{v_2}{v_1}$.

- Refractive index depends on the wavelength or frequency of light, $n(\lambda)$ or $n(\omega)$. Some examples of more accurate relation between the refractive index and wavelength of light is given, for example, by the Cauchy's and Sellmeier's equations.

- The refractive index $n(\omega)$ and dielectric function $\epsilon(\omega)$ are related by $n(\omega) = \sqrt{\epsilon(\omega)}$

- Some of the consequences and applications of refraction of light include: a parallel displacement of the emergent ray on passing through a rectangular block, decreased apparent depth of fluid as observed from air, total internal reflection and the critical angle, optical fiber, fish-eye-view of the world above water, displacement and bending of light in optical instruments, and the mirages in desert and at sea.

- Dispersion of light is the separation of white light in to its colored light components after refraction through a prism shaped refractive medium. Formation of the rainbow is due to the dispersion of light by the water droplets in the atmosphere after the rains.

- The refractive index of the material of the prism is given as: $\dfrac{n}{n_{air}} = \dfrac{\sin(\frac{A+D_{min}}{2})}{\sin(\frac{A}{2})}$

- Huygens wave theory explains the propagation of light as wavefronts, and can be applied to derive the laws of reflection and refraction.

- Finally Fermats principle states that light travels from one point to another point along the path of the least time of travel. It can be applied also to derive the laws of reflection and refraction.

2.7 Exercises

NOTE: Physical constants required to solve some of the exercises may be found either with the exercise or in the chapter.

2.1 A plane mirror is tilted at 30° above horizontal, and a ray is incident on it along the vertical direction. Calculate the angle of incidence, and the direction of the reflected ray from the vertical. Verify your answers from measurements from a ray diagram.

2.2 From the ray diagram in Figure 2.3, show that the image from a plane mirror is formed as far behind the mirror as the object is in the front. How does the image move when the object is moved:

(i) towards the mirror, *(ii)* away from the mirror?

2.3 From the ray diagram in Figure 2.3, show that the image from a plane mirror is the same size as the object.

2.4 The minimum length of a mirror required to see full image of a person is half the height of the person. Even then the clothing stores use full length mirrors in their dressing rooms, rather than economizing by using smaller length mirrors. Explain why?

2.5 A person uses a 10 cm long mirror placed at 25 cm from his eyes to see the full image of a high building located 50 m behind him. The image covers the entire length of the mirror. Draw a ray diagram to show the image formation as seen by the person, and calculate the height of the building.

2.6 Keeping the direction of an incident ray fixed, a plane mirror is rotated by an angle θ (Figure 2.5). Draw the ray diagram to show the reflected rays from both positions of the mirror, and calculate the angle between the two reflected rays.

2.7 Complete ray diagrams in Figs. 2.6, 2.7, and 2.8 to show the formation of images due to successive reflections, starting with the first reflection from the *other* mirror not shown in these ray diagrams.

2.8 Two plane mirrors are held edge to edge with their reflecting surface at θ^o, $(> 0^o and < 180^o)$. Draw a ray diagram to show two successive reflections of an incident ray from the mirrors, and calculate the angle by which an incident ray is rotated after the second reflection.

2.9 Calculate: (i) the speed of light, and (ii) relative refractive index with respect to air for each of the material given in Table 2.1. Arrange the materials in Table 2.1 in descending order of their optical density, and calculate the relative refractive index for each material with respect to the material just below in the 'descending' list.

2.10 Refractive indices given in Table 2.1 are for the mean wavelength of white light $(\lambda = 550\ nm)$. Calculate the range of refractive index for visible light $(\lambda_{violet} = 380\ nm, \lambda_{red} = 720\ nm)$ for each of the materials.

2.11 Calculate the angle of refraction when light traveling in air is incident at 30° at the air-material interface for each of the material given in Table 2.1. Calculate the angle by which the incident ray is rotated on entering the other medium. Repeat the calculations for the angle of incidence equal to 89.5^o.

2.12 Calculate the wavelength of light in each of the materials (except air) given in Table 2.1 for which the wavelength in air is 550 nm.

2.13 Light traveling in each of the medium except air in Table 2.1 is incident at 20° at the boundary of the medium with air. Calculate for each case the angle at which the ray emerges in air, and the angle by which the incident ray is rotated on entering air.

2.14 Prove that when a ray of light passes through two opposite sides of a glass block, the emergent ray is parallel to the incident ray. Prove that the lateral displacement of the emergent ray from the incident ray in terms of the thickness *(t)* of the block, angle of incidence *(i)*, and the refractive indices of air (n_a) and the material *(n)* of the block is given by:

$$d = t \sin i \left(1 - \frac{n_a \cos i}{\sqrt{n^2 - n_a^2 \sin^2 i}} \right)$$

2.15 A ray traveling in air is incident at 37° on a 5cm thick rectangular crown glass block, and emerges from the opposite face of the block. Calculate: (i) the time the ray takes to travel through the block, (ii) angle at which the ray emerges from the block, and (iii) the distance by which the emergent ray is displaced relative to the incident ray.

2.16 A container contains a 5 cm thick glass block $(n_g = 1.65)$ covered with 6 cm depth of water $(n_w = 1.33)$ followed by 4 cm thickness of oil $(n_o = 1.21)$ on top. Calculate the apparent depth of the bottom of the container as viewed from air $(n_a = 1)$.

Note: This exercise can be solved in two different but equivalent ways: *(i)* Since the depth of each material is seen from the air, calculate the apparent depth of each material directly as seen from air. *(ii)* Calculate the apparent depth of glass as seen from water, converting it to an equivalent depth of water. Next convert total water depth to an apparent depth with respect to oil giving a total effective oil depth. Finally, the total oil depth is converted to a total apparent depth as seen from air. Solve the problem in both ways, and compare the results.

2.17 A conically shaped diamond measures 2.5 *mm* from its top face to the conical tip. What shall be the apparent height of the diamond when viewed vertically through the top face?

2.18 A point source of light is placed at a depth of 2m in a swimming pool, and looking vertically down one sees a circle of light on the surface of the pool, Explain why one sees such a circle. Calculate: (i) the apparent depth of the source of light, and (ii) the diameter of the circle of light on the water surface.

2.19 Calculate the critical angle for each of the material given in Table 2.1. Compare these values with the angles of refraction corresponding to the angle of incidence equal to 89.5^o calculated in Exercise 2.11. Comment on any resemblance and differences between the two values for each material.

2.20 An optical fiber is made up of a plastic material of refractive index 1.45. The fiber is surrounded with air. Calculate the range of the angle from the cylindrical surface of the fiber for which the ray is propagated through the fiber without loss. To have this range as large as possible, should the refractive index be increased or decreased? Explain why, with an appropriate example. (Note: What happens when the refractive index becomes 1?)

2.21 A long cylindrical glass block is made of a material of refractive index 1.65. Rays of light enter the block from one of its circular face. Calculate the range of the angle of incidence of the rays which will be internally reflected from the cylindrical surface of the block.

2.22 A coin is placed at the center of the bottom of a 5 cm wide and 10 cm high cylindrical container. A person is looking into the container such that the top edge of the container is aligned with the diametrically opposite bottom corner. Water is gradually poured into the container. Calculate the depth of water when the coin becomes visible to the person. Explain qualitatively how the person should adjust his line of vision to see the coin when the depth of water is *(i)* less than, and *(ii)* larger than the one calculated for the original line of vision.

2.23 Arrange the materials in Table 2.1 in ascending order of their optical density, and calculate the critical angle for each material (except air) with respect to the material just above in the 'ascending' list.

2.24 A ray of light is incident at 30° on an equiangular prism of refractive index 1.5. Calculate (i) the angle of deviation of the emergent ray, (ii) the minimum angle of deviation for the ray from the same prism, and (iii) angle of incidence, angle of refraction in to the prism, and the angle of the emergent ray when the ray passes through the prism with minimum deviation.

2.25 Calculate the angles of minimum deviation for a ray passing through (i) equiangular prisms, and (ii) isosceles prisms of 45^0 angle made from (a) crown glass, (b) flint glass, (c) fused silica, and (d) diamond.

2.26 Refractive index of an equiangular glass prism is 1.5. Calculate the maximum angle of incidence on prism for which the ray shall emerge from the prism (not internally reflected at the other face of the prism). What is the angle of deviation for this extreme ray?

2.27 Angle of minimum deviation is measured using a hollow glass prism filled with water. What does this measurement represent, is it the angle of minimum deviation for water with respect to glass, or with respect of air or something else? Justify your answer with appropriate logic and reasoning.

2.28 Use Huygen's wave theory to prove that the angle of incidence is equal to the angle of reflection.

2.29 From Fermat's principle prove the law of reflection, *i.e.*, the angle of reflection is equal to the angle of incidence.

2.30 Consider rectangular axes in the plane on a beach with the *x*- axis parallel to the straight water-land boundary which is along the line $y = 5$. A lifeguard is locate at the origin of the axes, and a child is drowning at point *(10, 8)* , where all distances are in meters. The lifeguard's speeds on land and in water are $2\ ms^{-1}$ and $1.5\ ms^{-1}$ respectively. Find the optimum point of entry for the lifeguard in water so that the lifeguard can reach the victim in minimum time.

Chapter 3

Spherical Mirrors

3.1 Spherical mirrors and related terminologies

A spherical mirror is essentially a polished surface which is a portion of a sphere. There are two types of spherical mirors:

- concave mirror

- convex mirror

which are illustrated in Figures 3.1 (a) and 3.1 (b) respectively. A concave mirror is such that light is reflected from the inner surface, while for a convex mirror light is reflected from the outer surface of the sphere.

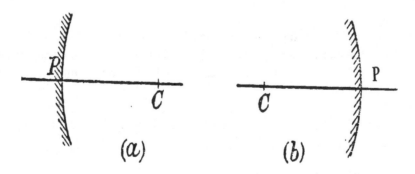

Figure 3.1: (a) A concave mirror is a portion of a sphere with center C such that its inner surface is reflecting. C is known as the center of curvature, P is the pole of the mirror, CP is the Principal axis. (b) A convex mirror is a portion of a sphere with center C such that its outer surface is reflecting. C is known as the center of curvature, P is the pole of the mirror, CP is the Principal axis.

There are various terminologies associated with spherical mirrors, for example: center of curvature, principal axis, radius of curvature, focal point and focal length. These terminologies are defined below, with labels referring to Figures 3.1 and 3.2.

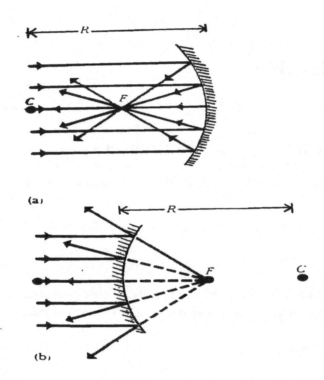

Figure 3.2: (a) Incident rays parallel to the principal axis after reflection *converge* at the focal point F of a concave mirror. R is the radius of curvature. (b) Incident rays parallel to the principal axis after reflection *diverge* as though emerging from the focal point F of a convex mirror. R is the radius of curvature.

The *center of curvature* C is a geometrical center of the sphere of which the spherical mirror is a portion.

The *radius of curvature* R is the radius of the sphere of which the spherical mirror is a portion.

The *pole* P is the physical center of the mirror located at the highest or lowest point on the mirrror surface. All distances in spherical mirrors optics are measured from P.

The *principal axis* is a straight line through the center of curvature and cuts the surface of the mirror at P.

The *focal point* F is a point through which all parallel rays incident on a concave mirror will be reflected through. For a convex mirror, the reflected rays appear to emerge from the point F and diverge.

The *focal length* (f) is half the radius of curvature, and by convention is positive ($f = R/2$) for a concave mirror and negative ($f = -R/2$) for a convex mirror. This is the distance from P to the focal point.

3.2 Image formation by spherical mirrors

Image formation by spherical mirrors is through reflection, unlike image formation by lenses where it is by refraction. We discuss the image formation by concave mirrors and by convex mirrors below, using ray diagrams. Ray diagrams are plotted using a suitable scale and complete characterization of the image is possible.

3.2.1 Image formation by concave mirrors

Image formation by concave mirrors can be studied by using ray diagrams. It is convenient to consider three rays only.

(1) A ray parallel to the principal axis is reflected back through the focal point.
(2) A ray through the focal point is reflected back parallel to the principal axis.
(3) A ray through the center of curvature is reflected back through the same path.

Applying the above rules, and referring to the rays as 1, 2 and 3 respectively, ray diagrams can be constructed with the object located in various positions infront of the concave mirror. The following notation is used: x_0 denotes the position of the object from the mirror, x_i denotes the position of the image from the mirror, and f is the focal length of the concave mirror. It should be mentioned that another common notation is to use u and v, which are related to x_0 and x_i, respectively, using the following prescription:

$$x_0 \leftrightarrow u$$
$$x_i \leftrightarrow v \qquad\qquad (3.1)$$

(a) $x_0 > 2f$: The characteristics of the image are that it is real, inverted and reduced in size, and is located between C and F.

(b) $x_0 = 2f$: The characteristics of the image are that it is real, inverted and of the same size as the object, and is located at C.

(c) $f < x_0 < 2f$: The characteristics of the image are that it is real, inverted and enlarged in size, and is located beyond 2f.

(d) $x_0 = f$: The characteristics of the image are that it is real, inverted and enlarged in size, and is located at infinity.

(e) $x_0 < f$: The characteristics of the image are that it is virtual, upright and enlarged in size, and is located behind the mirror.

The respective ray diagrams are illustrated in Figure 3.3. The sixth case for which the object is located at $x_0 \to \infty$, is left as an exercise for the student. What are the characteristics of the image? Where is it located?

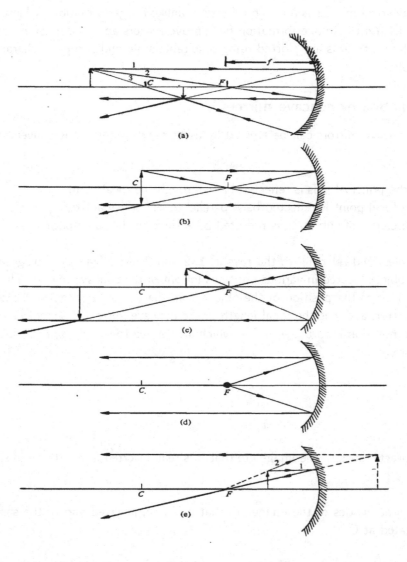

Figure 3.3: Image formation by a concave mirror with the object located in various positions: (a) $x_0 > 2f$, (b) $x_0 = 2f$, (c) $f < x_0 < 2f$, (d) $x_0 = f$ and (e) $x_0 < f$.

3.2.2 Image formation by convex mirrors

Ray diagrams are also used to study image formation by convex mirrors . The same type of three rays as used in the case for conncave mirrors can be used, but the ray parallel to the principal axis is now observed to *diverge* from the focal point. The ray diagram for image formation by a convex mirror is illustrated in figure 3.4. A convex mirror always produces a virtual image irrespective of the postion of the object. There is no need of considering six cases as was done for the concave mirror. The characteristics of the image formed by a convex mirror are that, irrespective of the object location, the image is always:

(i) virtual
(ii) upright
(iii) reduced in size
and lies between the focal point and the mirror.

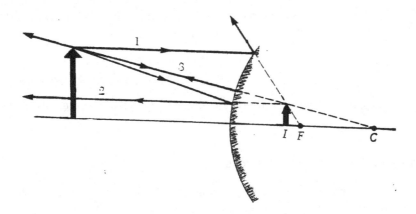

Figure 3.4: Image formation by a convex mirror. A convex mirror always produces a virtual, upright and reduced image at I irrespective of object location.

3.3 Sign convention and derivation of the mirror equation

3.3.1 Sign conventions for spherical mirrors

In solving optical problems using spherical mirrors certain sign conventions have been adopted. These are as follows:

For *concave mirrors* the sign conventions are as follows:

(i) The position of the object, $x_0 = \begin{cases} \text{Positive for a real object} \\ \text{Negative for cases of a virtual object} \end{cases}$

(ii) The focal length, f, is positive.

(iii) The position of the image, $x_i = \begin{cases} \text{Positive for a real image} \\ \text{Negative for a virtual image} \end{cases}$

(iv) The magnification, $m = h_i/h_0 = -x_i/x_0 = \begin{cases} \text{Positive for an upright image} \\ \text{Negative for an inverted image} \\ \text{is such that} |m| > 1 \text{ for an enlarged image} \\ \text{is such that} |m| < 1 \text{ for a reduced image} \end{cases}$

where h_0 and h_i are the heights of the object and image, respectively.

For *convex mirrors* the sign conventions are as follows:

(i) The position of the object, $u = \begin{cases} \text{Positive for a real object} \\ \text{Negative for a virtual object} \end{cases}$

(ii) The focal length, f, is negative.

(iii) The position of the image, $x_i = \begin{cases} \text{Positive for a real image, for combined mirrors and/or lenses} \\ \text{Negative for a virtual image} \end{cases}$

(iv) The magnification, $m = h_i/h_0 = -x_i/x_0 = \begin{cases} \text{Positive, always} \\ \text{is such that} |m| < 1 \text{ for a reduced image, always} \end{cases}$

Now, the above rules are essential and very useful in solving geometrical optics problems involving spherical mirrors.

3.3.2   Mirror equation for concave mirrors

To prove the mirror equation for concave mirrors, we consider one of the typical image formation configurations, say as illustrated in Figure 3.5, where u denotes the object distance from the mirror, v denotes the image distance from the mirror, and f is the focal length of the concave mirror. Note that $x_0 \leftrightarrow u$ and $x_i \leftrightarrow v$.

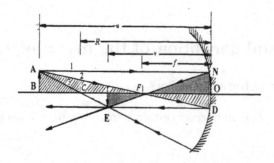

Figure 3.5: Typical image formation by a concave mirror.

It can be noted that $\triangle ABF$ and $\triangle DOF$ are similar. Hence

$$\frac{AB}{DO} = \frac{BF}{OF}$$

from which, noting that $h_0 = AB = NO$ is the height of the object, and $h_i = DO = EI$ is the height of the image, one obtains

$$\frac{h_0}{h_i} = \frac{x_0 - f}{f}$$

or

$$\frac{h_i}{h_0} = \frac{f}{x_0 - f} \tag{3.2}$$

Also, $\triangle IEF$ and $\triangle ONF$ are similar. Hence

$$\frac{IE}{ON} = \frac{IF}{OF}$$

from which, one obtains

$$\frac{h_i}{h_0} = \frac{x_i - f}{f} \tag{3.3}$$

Equating equations (3.2) and (3.3), and cross multiplying, one obtains

$$f^2 = (x_0 - f)(x_i - f)$$

which leads to

$$\frac{1}{x_0} + \frac{1}{x_i} = \frac{1}{f} \tag{3.4}$$

known as *the mirror equation*. As we shall see in the next section, the structure of the mirror equation for the convex mirror is similar to this one.

The magnification by a mirror is given by

$$m = -\frac{x_i}{x_o} = \frac{h_i}{h_o}$$

3.3.3   Mirror equation for convex mirrors

To prove the mirror equation for convex mirrors , we consider the typical image formation configuration, say as illustrated in Figure 3.6. The same notation for x_0, x_i and f is used, noting that x_0 denotes the object distance from the mirror, $-x_i$ denotes the image distance from the mirror. *Note that $-x_i$ is positive due to sign convention of a virtual image, and $-f$ is the focal length of the concave miror, since $-f$ is positive due to sign convention of the focal length of a convex mirror.*

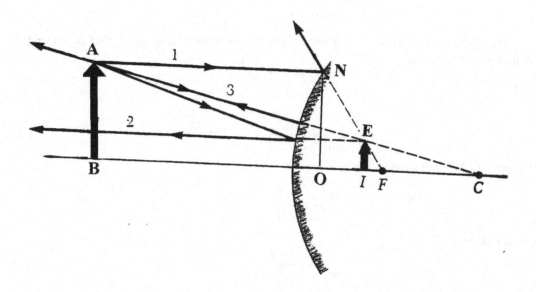

Figure 3.6: Typical image formation by a convex mirror.

It can be noted that $\triangle ABC$ and $\triangle EIC$ are similar. Hence

$$\frac{EI}{AB} = \frac{IC}{BC}$$

from which, noting that $h_0 = AB = NO$ is the height of the object, and $h_i = DO = EI$ is the height of the image, one obtains

$$\frac{h_i}{h_0} = \frac{-2f + x_i}{x_0 - 2f} \tag{3.5}$$

Also, $\triangle NOF$ and $\triangle EIF$ are similar. Hence

$$\frac{EI}{NO} = \frac{IF}{OF}$$

from which, one obtains

$$\frac{h_i}{h_0} = \frac{-f + x_i}{-f} \tag{3.6}$$

Equating equations (3.5) and (3.6), and cross multiplying, one obtains

$$-f(-2f + x_i) = (x_0 - 2f)(-f + x_i)$$

which leads to

$$\frac{1}{x_0} + \frac{1}{x_i} = \frac{1}{f}, \tag{3.7}$$

and as before, this is *the mirror equation*.

3.4 Discussion using graphs

3.4.1 Graph of $1/x_i$ against $1/x_0$

The mirror equation proved earlier can be re-arranged and written in the form:

$$\frac{1}{x_i} = -\frac{1}{x_0} + \frac{1}{f} \qquad (3.8)$$

from which one can note that a graph of $1/x_i$ against $1/x_0$ will be a straight line with a slope of -1 (negative slope). The focal length is extracted as the reciprocal of the intercepts. For example, with the graph in Figure 3.7, the focal length, $f = 10.0$ cm.

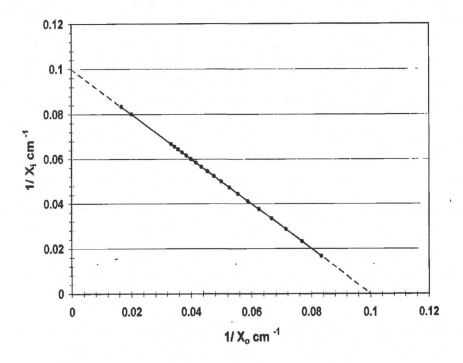

Figure 3.7: A graph of $1/x_i$ against $1/x_0$ for a concave mirror, giving a the focal length of $f = 10.0$ cm from the reciprocal of the intercept along the $1/x_i$ axis.

3.4.2 Graph of m against x_0/f

The magnification, $m = -x_i/x_0$. Using the mirror equation, m can be written in the form

$$m = \frac{1}{1 - \frac{x_0}{f}} \qquad (3.9)$$

A graph of m against x_0/f for both the concave mirror and the convex mirror is illustrated in Figure 3.8. The image characteristics corresponding to several image positions are illustrated in Table 3.1. The graph and the table are related by comparing the image characteristics in the various regions.

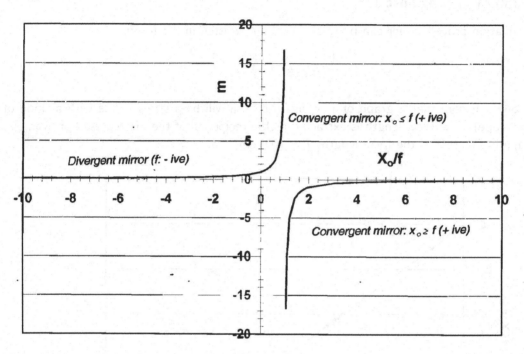

Figure 3.8: A graph of m against x_0/f for a convex (divergent) mirror and concave (convergent) mirror.

Table 3.1: Image characteristics of an object placed before a convex mirror and concave mirror.

Value of $\frac{x_0}{f}$	Value of m	Image characteristics
Concave mirror		
$0 < \frac{x_0}{f} < 1$	$m > 1$	virtual, upright, enlarged
$\frac{x_0}{f} = 1$	$m = \infty$	virtual, upright, enlarged(*)
	$m = -\infty$	real, inverted, enlarged(*)
$1 < \frac{x_0}{f} < 2$	$m < -1$	real, inverted, enlarged
$\frac{x_0}{f} = 2$	$m = -1$	real, inverted, same size
$\frac{x_0}{f} > 2$	$-1 < m < 0$	real, inverted, diminished
Convex mirror		
Any position	$0 < m < 1$	virtual, upright, diminished

(*) See discussion of the analogous case for lenses in connection with Figure 4.8.

3.5 Chapter 3 Summary

The Chapter deals with properties of, and image formation from spherical mirrors that are sections of a spherical shells and are of two types: Concave and Convex. They are also known as convergent and divergent mirrors respectively. For concave mirror the inner surface of the shell is reflecting, and for convex mirror the outer surface of the shell is reflecting. The mirrors have a circular cross section, diameter of which is known as the aperture of the mirror.

- The center of curvature, C, and the radius of curvature, r, of the mirror refer to the corresponding center and radius of the spherical shell of which the mirror is a section. It is obvious to note that C lies in front of the reflecting surface of the concave mirror and is treated as a real point, and it lies behind the reflecting surface for the convex mirror and is treated as a virtual point. Line joining C to any point on the surface of the mirror is normal to the surface. Why?

- Pole, P, is the physical center of the mirror. It is the deepest point on the surface of the concave mirror, and the highest point on the surface of the convex mirror. All distances for image characterization are measured from P. Line joining P to C (and extended both ways) is the principal axis of the mirror.

- Reflection from spherical mirrors follow the usual laws of reflection as discussed in Chapter 2, but reflection of three rays is of particular interest. Any two of these rays can be used to locate and characterize the image by geometrical method from a ray diagram drawn to scale. Different scales are used for distances along the principal axis (x-axis) and lengths transverse to the principle axis (y-axis). A common mistake by students is to use the same scale for both.
  o Ray 1 incident parallel to the principle axis on reflection from a concave mirror passes through (converges to) a point, F, between C and P on the principle axis, and on reflection from a convex mirror appears to diverge from point F between C and P on the principle axis. F is known as the focal point of the mirror. It is a real point for the concave mirror and a virtual point for the convex mirror. Distance PF is the focal length, f, of the mirror. For mirrors with small aperture, (compared to the radius of curvature), $f = (r/2)$.
  o Incident Ray 2 passing through C (concave mirror) or incident in the direction of C (convex mirror) is reflected back along the same direction. Why?
  o Incident Ray 3 passing through F (concave mirror) or incident in the direction of F (convex mirror) is reflected parallel to the principle axis. This, together with Ray 1 illustrates the reversibility of the path of light.

- Six cases of image formation from a concave mirror are of interest: (i) $x_o = \infty$: Image is real, inverted, reduced in size and lies at distance $f(F)$. (ii) $\infty > x_o > 2f$: Image is real, inverted, reduced in size and lies between distance $2f(C)$ and $f(F)$. (iii) $x_0 = 2f$: Image is real, inverted, same size as the object and lies at distance $2f(C)$. (iv) $2f > x_o > f$: Image is real, inverted, enlarged in size and lies beyond $2f(C)$ distance. (v) $x_o = f$: Image is real, inverted, enlarged in size and lies at ∞. (vi) $x_o < f$: Image is virtual upright, enlarged in size and lies behind the mirror.

- For a convex mirror there is only one case of image formation. For any location of the object, the image is virtual, upright, reduced in size and lies behind the mirror between the focal point F and pole P.

- Analytically one uses the mirror equation; $(1/xo) + (1/xi) = (1/f)$, and the expression for magnification: $m = (hi/ho) = (-)(xi/xo)$ in conjunction with a sigh convention to locate and characterize the image.

- Sign Convention: *All real distances are positive, and all virtual distances are negative.*
  ○ *f and r are (+) for a concave mirror, and (-) for the convex mirror.*
  ○ x_o for a single mirror is always (+), but for a combination of mirrors/ mirrors and lenses it could be (-) if the intermediate image is virtual.
  ○ x_i is (+) for cases (i) to (v) of the concave mirror, and (-) for case (vi) of the concave mirror and the convex mirror. For a combination of mirrors/ mirrors and lenses a care must be taken to choose the correct sign.
  ○ m is (+) for an upright and virtual image, and (-) for an inverted and real image.
  ○ When the known quantities are used with a correct sign, the calculated quantity comes out with the correspondingly correct sign, which is a cross-check for the correctness of signs used in calculation.

3.6 Exercises

3.1. (a) An object is placed 5 cm in front of a concave mirror with a radius of curvature of 20 cm. Calculate the image position and the magnification and characterize the image.
(b) Solve the above problem by ray diagram and compare the results with part (a).

3.2. (a) An object of size 3.0cm is placed 50 cm in front of a convex mirror whose focal length has a magnitude of 40 cm. Calculate the image position, size of the image and characterize it.
(b) Solve the above problem by ray diagram and compare the results with part (a).

3.3. A dentist holds a mirror 2 cm from a tooth, and sees an upright image which is magnified four times.
(a) State, with reasons, what type of mirror is used, and calculate its focal length.
(b) Illustrate with a sketch ray diagram the above observation.

3.4.(a) An image of a matchstick is formed 10 cm behind a convex mirror whose radius of curvature is of magnitude 30 cm. Calculate the position of the matchstick from the mirror.
(b) Assuming the size of the matchstick to be the one usually found in most of Botswana stores, calculate the size of the image, and characterize it.

3.5. An image formed by a a spherical mirror whose magnitude of the focal length is f is found to be one-half the size of the object.
(a) If the mirror is concave, find the object position in terms of f, and characterize the image.

(b) For the case of a convex mirror, find the object position in terms of f, and characterize the image.

3.6. Draw ray diagrams for a ray parallel to the principal axis reflected from a concave mirror and a convex mirror. Show also the normals at the point of incidence for both cases. Prove that $f = |\frac{R}{2}|$ for both the cases. State any assumptions made, and from this exercise what can you state about the focal length of mirrors with large aperture?

3.7. An object is located at 2.4 m distance from a wall. Using a spherical mirror, an image of the object magnified three times is projected on the wall. State the type of the mirror used giving reason for your answer. Calculate the radius of curvature of the mirror, and its distance from the wall.

3.8. Use ray diagrams 3.3 (c) and 3.3(e) to derive the mirror equation.

3.9. A graduated one meter long stick is placed along the principal axis of a concave mirror of focal length 40 cm such that the zero mark of the stick is nearer to the mirror at a distance of 60 cm. The meter stick is very slightly misaligned so that rays from all parts of the stick can reach the mirror, and get reflected. Calculate the location of the images of the two ends of the meter stick. Qualitatively sketch the image along the principle axis, clearly showing the positions of the 0, 20 cm, and 100 cm marks in the image. Can you draw some conclusion about how the two regions of space divided by the center of curvature are imaged by a concave mirror?

3.10. Repeat Exercise 3.9 for a convex mirror of focal length 40 cm.

3.11. A thin spherical mirror is made in such a way that its both surfaces are reflecting. When it is used as a concave mirror, an inverted image of an object placed at xcm is formed at 1.2 m from the mirror. Keeping the object distance the same, when convex side of the mirror is used, an upright image is formed at 17.4 cm. Calculate focal length for both the mirrors. State if you have made any assumptions.

3.12. When an object is placed at a distance U_1 from a concave mirror, a real image of magnification 3 is formed. When the object is placed at a distance U_2, a virtual image three times the object size is formed. (i) Calculate the ratio $|U_1/U_2|$. (ii) If the separation between the two positions of the object is 40 cm, calculate the focal length of the mirror, and the distance of the object for the two cases.

3.13. Plot qualitatively a graph of (x_i/f) against (x_o/f) for the convergent and the divergent mirrors on a single diagram. Use the diagram to discuss how the image is displaced as the object is moved from a very large distance $(x_o \rightarrow \infty)$ to very close to the mirror for both types of mirrors.

3.14. A concave and a convex mirror both of focal length 20 cm each are placed 40 cm apart with their reflecting surfaces facing each other. An object is placed between the two mirrors at 30 cm from the concave mirror. Locate and characterize the image formed after one reflection from each mirror for the following two cases. (i) First reflection from the concave mirror, and second reflection from the convex mirror. (ii) First reflection from the convex mirror and second reflection from the concave mirror.

3.15. A plane mirror and a concave mirror of focal length 20 cm are placed 100 cm apart facing each other. An object is placed between them at 60 cm from the concave mirror. Locate and characterize the image formed after one reflection from each mirror for the following two cases. *(i)* First reflection from the concave mirror and the second reflection from the plane mirror. *(ii)* First reflection from the plane mirror and the second reflection from the concave mirror.

3.16. Repeat Exercise 3.15 for a convex and a plane mirror combination, all distances remaining the same.

Chapter 4

Thin Spherical Lenses

A *lens* is an optical component made from a refractive medium, bounded by two surfaces at least one of which is a curved surface. A lens is used to converge or diverge a beam of light by refraction that produces a reduced or enlarged, real or virtual, and inverted or upright image of an object placed on one side of the lens. The image may lie on the same side as the object or on the other side of the lens. The characteristics of the image and its location depend on the type of the lens, and on the distance of the object from the lens.

It was shown in Chapter 2 that a ray of light after refraction through a prism, bends towards its base. The principle of the working of a lens can be demonstrated by using a pair of prisms as in Figure 4.1.

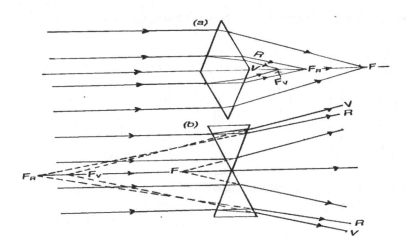

Figure 4.1: A pair of prisms used to (a) converge and (b) diverge a parallel beam of light demonstrates the principle of convergent and divergent lenses respectively.

When a parallel beam of light passes through two prisms placed base-to-base, the emergent rays bend towards the common base line and converge to points on the line (Figure 4.1 a). On the other hand if the prisms are placed apex-to-apex, the emergent rays bend away from the apex (towards the base) of the respective prism so that the beam diverges and appears to come from virtual points on the line through the common apex and parallel to the bases of the prisms (Figure 4.1 b). These two combinations of a pair of prisms may be visualized as the converging and diverging lenses respectively. However, a lens made from a pair of prisms does not have a sharply defined point of convergence or divergence (focus), and if white light is used, lights of different colours would have different focal points due to the dispersion of light. Thus the image formed from a pair of prisms can not be sharply focused, and has colour defect. These two defects in the case of lenses are known as spherical and chromatic aberrations, and are discussed later in section 4.5. Here it suffices to state that both these defects which are very prominent in the 'pair of prisms lens' are greatly reduced by modifying the refracting surfaces of the combination of prisms to continuous, smooth curved surfaces. Dictated by the requirement of application, curved surface(s) of a lens may be spherical, cylindrical, parabolic or hyperbolic that give a spherical, a cylindrical, a parabolic, or a hyperbolic lens respectively. There is yet another type of lens, known as the Fresnel lens for which one of the surface is zagged-step shaped. In this chapter we shall concern ourselves with the optics and applications of thin spherical lenses only.

4.1 Spherical Lenses: Definitions and Terminology

For the spherical lenses at least one refracting surface is spherical, and the other surface may be spherical or plane. Some possible combinations of spherical and plane surfaces for spherical lenses, their names, and their converging/ diverging features are given in Figure 4.2.

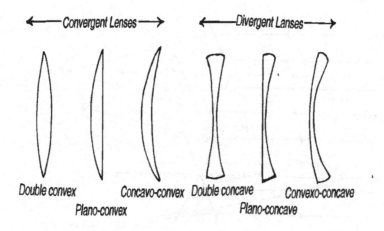

Figure 4.2: Various types of spherical lenses.

Spherical surface of a lens may be viewed as a segment of a sphere of radius r as shown in Figure 4.3 for a double convex and a double concave lens. Various terms that apply to lenses are defined using Figure 4.3.

Centers and radii of curvature are the centers and radii of the two spheres of which the two surfaces of the lens are segments. Note how the locations of the two centers of curvature C_1 and C_2, and the radii of curvature r_1 and r_2 for two surfaces of a bi-convex and a bi-concave lens are switched with respect to the lens. A plane surface may be regarded as a segment of a spherical surface with an infinite radius of curvature for which the center of curvature lies at infinity. A line joining the center of curvature to a point on the corresponding spherical surface (radius vector)is normal to the surface at the point. Line C_1C_2 joining the two centers of curvature is the *principal axis* of the lens. A ray along the principal axis is refracted through the lens undeviated (Why?).

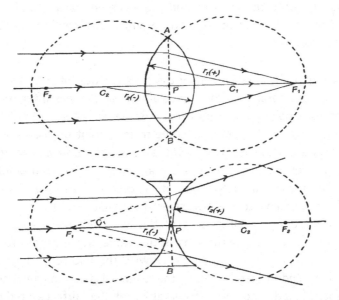

Figure 4.3: Definitions and terminology for double convex and double concave lenses.

A point P on the principal axis which lies either inside the lens or very close to it on the outside (Figure 4.4) is the *optical center* of the lens. A ray passing through the optical center is also refracted through the lens undeviated. All distances in a lens are measured from the optical center.

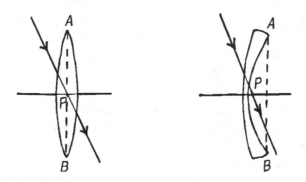

Figure 4.4: Optical center of a lens may lie inside or outside the thin lens.

The rays through the optical centers for two cases shown in Figure 4.4 are a simplified representation which can be applied to the ray diagrams for thin lenses only without much loss in accuracy. Under real conditions an incident ray on entering the lens bends towards the normal, then it passes through the optical center, and on emerging into the surrounding medium (air) from the other surface it bends away from the normal. If the two surfaces of the lens are symmetrical, the incident and the emergent rays are parallel to each other, and the separation between them depends on lens thickness. For a thin lens the separation between the incident and the emergent rays is small, and within first approximation the two rays can be treated to be along a single straight line passing through the optical center as shown in the figure.

Aperture of a lens is the diameter *AB* (Figure 4.3) of the cross section of the lens which intercepts the incident light. A beam of light parallel to the principal axis of the lens after refraction through a convergent lens converges to a point on the principal axis on the other side of the lens, and in the case of a divergent lens diverges and appears to come from a point on the principal axis on the same side from which the rays are incident. This point is called the *principal focus* or focal point F_1 of the lens. The focal point for a convergent lens is real, whereas the focal point of a divergent lens is virtual. Distance of the focal point from the optical center, $PF_1 = f_1$, is the *focal length* of the lens (Figure 4.3). Corresponding to the two sides from which light can incident on a lens, there are two focal points F_1 and F_2 on either side of the lens, and there are two focal lengths f_1 and f_2. If the medium on both sides of the lens is the same, for example for a lens used in air, both focal lengths are equal, *i.e.*, $f_1 = f_2 = f$, otherwise the two focal lengths are different. Planes through the focal points and perpendicular to the principal axis are called the *focal planes*. Relationship between the focal length and radii of curvature of the surfaces of the lens is not as simple as it is for the spherical mirrors, and the focal length also depends on the refractive indices of the material of the lens and the medium in which it is used. This relationship is derived and discussed in a later section. Furthermore, unlike the spherical mirrors, centers of curvature of a lens do not have any special significance as far as the ray diagrams and image formation are concerned, except that the radius vector gives the direction of the normal to the refracting surface. Instead, for lenses one uses the distance equal to *2f* which plays the same role as the radius of curvature *(= 2f)* for the spherical mirrors. The *power of a lens P* is defined as the reciprocal of its focal length in meters, *i.e.*, $P = (1/f(m))$. The units of power is *Diopter (D)*. A lens with small focal length causes larger convergence or divergence of parallel rays because the focal point lies closer to the lens, and the lens is said to be more powerful, *i.e.*, *P = (1/f)* is large. The *lateral (or linear) magnification (m)* is defined as the ratio of the image height (h_i) to the object height h_o, *i.e.* $m = h_i/h_o$.

In this Chapter we shall deal with refraction from, and characteristics of images formed from *thin* or *small* lenses only. Although, more insight into a thin/ small lens shall be gained later when we discuss the defects of lenses. At this stage we can define a *thin/small lens* as having a small aperture compared to its focal length. Practical consequence of this is that, a thin/ small lens has a well defined, not necessarily a point-sharp, focal point. A lens with an aperture about $1/5^{th}$ of its focal length can be regarded as a thin lens. In fact thinness or smallness of a lens is relative, and thinner or smaller is the lens, sharper is the focal point. On the other hand, *thick* or a large lens has a diffused focal point spread over a range along the principal axis.

4.2   Refraction from Lenses: Geometrical Approach

In this section we develop, discuss and apply the rules to draw ray diagrams to locate and characterize images formed by lenses. Passage of a ray of light through a lens is governed by the laws of refraction at both the surfaces of the lens. When a lens is used in air, the incident ray traveling from the rarer to denser medium bends towards the normal, and the emergent ray traveling from denser to rarer medium bends away from the normal. The normals to the refracting surfaces are given by the radius vectors at the points of refraction. However, these rules can not be used conveniently to draw ray diagrams to locate and characterize images formed by a lens. Instead, from the definitions of the focal points and the optical center we identify three different rays, any two of which, can be used to draw the ray diagrams to locate and characterize the image formed from a lens(Figure 4.5).

- A rays incident parallel to the principal axis (ray 1), after refraction through a lens passes through the focal point (convergent lens), or appear to diverge from the focal point (divergent lens).

- A ray passing through the focal point on the incidence side of a convergent lens, or incident in the direction of the focal point on the other side of a divergent lens, after refraction through the lens goes parallel to the principal axis (ray 2).

- A ray passing through the optical center of the lens (ray 3) passes through the lens undeviated.

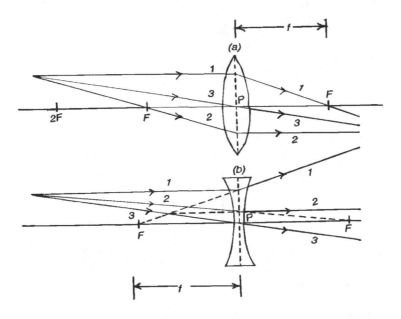

Figure 4.5: Any two of the three rays, 1, 2 and 3, can be used to locate and characterize images formed by (a) convergent, and (b) divergent lenses.

4.2.1 Location and characterization of images from ray diagrams

One can locate and characterize the image of an object placed in front of a lens by plotting a to-the-scale ray diagram. A common mistake made by students while drawing to-the-scale ray diagrams is that they tend to use the same scale for the distances from the lens and for the heights of the object and the image. One must use different, and convenient scales for the distances and the heights, and for greater accuracy of measurements scales should also be the largest possible so that the ray diagram fills most of the available space. The convenient scale implies that it should be possible to apply the scale to fractional distances with minimal need to estimate, for example 3 or 7 units of a measured distance equated to 1 or 2 units of scale is not a convenient scale, while 1, 2, 4, 5, or 10 etc units of measured distance equated to 1 unit of scale are convenient scales. A lens is represented by a vertical line through the optical center perpendicular to the principal axis, and convergent and divergent lenses are distinguished by converging or diverging arrow heads respectively at the two ends of the line, as shown in the ray diagrams that follow.

Ray diagrams for a convergent lens

There are six distinct cases of ray diagrams for a convergent lens. These cases depend on the distance of the object from the lens which may vary from ∞ to less than the focal length f of the lens. Ray diagrams drawn using two of the three rays specified in Section 2., and image characteristics for these cases follow.

Case (i) $x_0 = \infty$

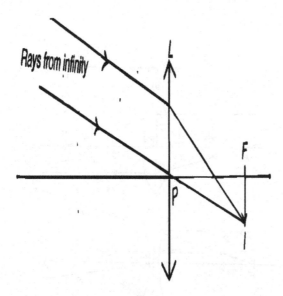

Figure 4.6a: Object is placed at infinity.

Image is real, inverted, smaller than the object, and lies in the focal plane (on the other side) of the lens. This case represents a simple telescope.

Case (ii) $x_0 > 2f$

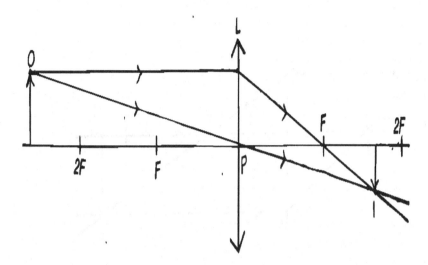

Figure 4.6b: Object is placed between infinity and *2f* distance from the lens.

Image is real, inverted, smaller than the object, and is located between *f* and *2f* distance (on the other side) from the lens. This case represents a simple photographic camera.

Case (iii) $x_0 = 2f$

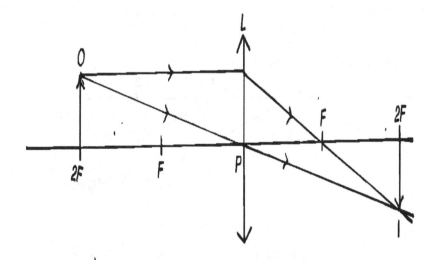

Figure 4.6c. Object is placed at *2f* distance from the lens.

Image is real, inverted, same size as the object, and is located at *2f* distance (on the other side) from the lens. This represents the function of a simple photocopier.

Case (iv) $f < x_0 < 2f$

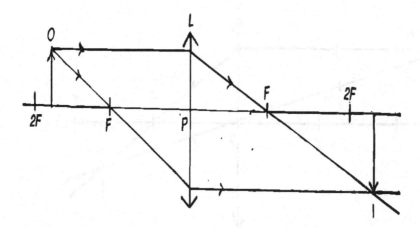

Figure 4.6d: Object is placed between *2f* and *f* distance from the lens.

Image is real, inverted, larger than the object, and is located between *2f* and *f* distance (on the other side) from the lens. This represents a basic slide projector.

Case (v) $x_0 = f$

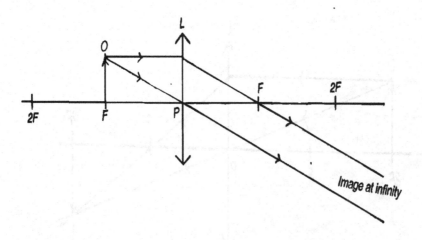

Figure 4.6e: Object is placed in the focal plane (at distance *f*) of the lens.

Image is real, inverted, larger than the object, and is located at infinite distance (on the other side) from the lens. This case represents the principle of flood/ security lights.

Case (vi) $x_0 < f$

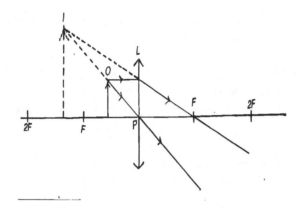

Figure 4.6f: Object is placed at a distance less than f from the lens.

Image is virtual, upright, larger than the object, and is located beyond f (on the same side as the object) from the lens. This is the case of a simple microscope or a magnifier, for example the watch makers lens.

Ray diagram for a divergent lens

For a divergent lens location and characteristics of the image do not depend on the distance of the object from the lens. There is only one case of ray diagram as shown in Figure 4.7. The image is virtual, upright, smaller than the object, and lies between the object and the lens, at a distance less than f.

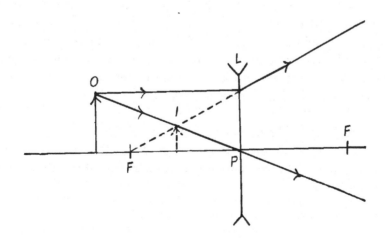

Figure 4.7: The only case of ray diagram for a divergent lens.

General Characteristics of the Images

From the above 7 cases of ray diagrams, the following interesting general features of images formed by lenses may be noted.

1. Real images are always inverted with respect to the object, and are on the other side of the lens, whereas the virtual images are upright with respect to the object, and lie on the same side of the lens as the object. Location of real and virtual images with respect to the lens is a rather obvious conclusion. Why?

2. For a convergent lens location of the image and its characteristics depend on the distance of the object from the lens but this is not so for the divergent lens.

3. For object distance $\geq f$, image formed by a convergent lens is real and inverted, and for distance $<f$ the image is virtual, upright and enlarged. On the other hand for divergent lens the image is always virtual, upright and smaller irrespective of the distance of the object from the lens.

4. For a convergent lens as the object approaches the lens from infinity to *2f*, the image moves from *f* to *2f*, and its size increases from very small to the same size as the object. Thus the entire distance from infinity to *2f* from the lens is mapped as an image on to a very short distance from *f* to *2f*. A large displacement of object causes a relatively very small displacement of the image.

5. Again for a convergent lens as the object approaches the lens from *2f* to *f*, the image moves from *2f* to infinity, and its size increases from the same size as the object to a very large size. Thus the short distance from *2f* to *f* from the lens is mapped as an image on to a very large distance from *2f* to infinity. A small displacement of object causes a relatively very large displacement of the image.

6. The above two observations for the images from a convergent lens have an important bearing on the experimental determination of the focal length of a convergent lens from the measurements of object and image distances. If the object is placed beyond *2f*, the uncertainty in the measurement of the object distance is large, and the uncertainty in the measurement of the image distance is small. On the other hand if the object is placed at less than *2f* distance then the uncertainty in the object distance is small and the uncertainty in the image distance is large.

7. For a convergent lens, cases *(i)* and *(ii)* are complimentary to cases *(v)* and *(iv)* cases respectively with *(iii)* as the central case, *i.e.*, if one changed the image to object then the object becomes the image and vice versa. This re-enforces the principle of the reversibility of the path of light discussed elsewhere in the book.

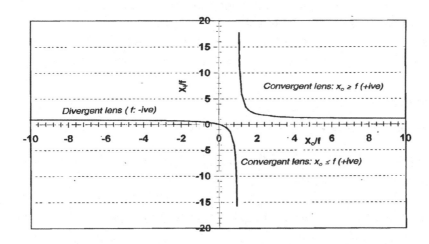

Figure 4.8: Normalized image distance as a function of the normalized object distance for lenses.

Figure 4.8 shows normalized image distance as a function of normalized object distance (in unit of the focal length) for both types of lenses. The figure generated using analytical relation (discussed in section 4.4.1) between the object and image distances and the focal length of the lens summarizes graphically some of the above conclusions. It is left as an exercise for students to qualitatively draw the same graph from to-the-scale ray diagrams. From this figure an ambiguous but interesting situation for the image of an object placed at a distance equal to f from a convergent lens warrants discussion. When one approaches the focal point from infinity, the image of an object placed at f is real, located at infinity. On the other hand if one approached the focal point from the lens, the image of an object placed at f ($x_0/f = 1$) shall be virtual and at infinity on the same side as the object. This ambiguity can be resolved in terms of a parallel beam of rays obtained after refraction from the lens. Such a beam may either be regarded as converging to a point at infinity so that a real image at infinity is formed, or may be regarded as diverging from a point at infinity which shall give a virtual image at infinity. However, we know from experience that the image of an object at infinity (very large distance such as the sun) is real and lies at the focal point. Therefore, from the reversibility of the object and image, the image of an object placed at the focal point should be real located at infinity. But this is not the complete picture, and there is yet another dimension to the problem. Due to a defect of spherical lenses known as spherical abberations discussed in section 4.5.1, spherical lenses no matter how small and thin they are, do not have a sharply defined focal point. The focus lies within a small but finite spread along the principal axis, and the cases of an object and the image at the focal point are a theoretical ideality. They do not apply to real spherical lenses as precisely as assumed for an ideal perfect lens.

4.3   Refraction at Spherical Surfaces: Analytical Approach

In this section analytical relations for refraction at spherical surfaces separating two media, and for refraction through lenses are derived. In order to use these relations to locate and characterize images analytically, one must use a sign convention for various distances. We discuss the sign convention first.

4.3.1   Sign Convention

Analytical expressions for lenses involve the following distances:

(i) The object and the image distances from the lens, and the focal length of the lens which are measured from the optical center along the principal axis,

(ii) heights of the object and the image measured perpendicular to the principal axis, and

(iii) radii of curvature of the lens surfaces.

A specific sign convention is used to express these distances and heights. In addition, there is a sign convention for the magnification *m*

Sign convention for distances measured along the principal axis, can be stated in brief as follows: *'the real distances are positive, and the virtual distances are negative.'* This in essence means that distances to real object, real image and real focal point are positive, and distances to virtual object, virtual image and virtual focal point are negative. This sign convention implies:

- Focal length of a convergent lens is always positive and that for a divergent lens is always negative. Why?

- When only one lens is used, the object is invariably real, and consequently the object distance is positive. For example for all the seven cases of ray diagrams in section 4.2.1 the object distance is positive. However, when a combination of two or more lenses or lenses and mirrors is used, the image formed from one component acts as an object for the next component in the sequence. Such an object could be real or virtual, and the corresponding object distance would be positive or negative respectively. One needs to be rather careful in dealing with the sign of such 'objects', because an intermediate image may itself be virtual but it may act as a real object for the next component or vice versa. For example consider two scenarios. In the first case a divergent lens is placed to the left of another lens (or a mirror), and the object is placed to the extreme left. The first image formed by the divergent lens is virtual, but it shall act as a real object for the component to the right of the divergent lens. For the second case consider a convergent lens placed to the left of another lens or mirror, and the two are very close together, at a distance $< f$ of the convergent lens. The object is placed to the extreme left at a distance $> f$ from the lens. The image formed by the convergent lens falls to the right of the lens and is real. However, if the second component is placed at such a distance from the convergent lens that it intercepts the refracted rays from the lens before the image is formed, then no intermediate image, real or virtual, would be present. Therefore, the nonexistent, 'real', intermediate image from the convergent lens acts as a virtual object for the second component. A simple rule to decide the sign for the object distance in such cases is: *"If the object distance is measured opposite to the direction of incident rays then the object is treated as real and the object distance is positive, If the distance is measured in the same direction as the incident rays then the object is treated as virtual and the object distance is negative"*

- Sign of the image distance when only one lens is used is also straight forward. For example again considering the ray diagrams in section 4.2.1, images for the first five cases of convergent lens are real, and their distances from the lens are positive. For the sixth case of the convergent lens, and for the divergent lens the images are virtual and their distances are negative. However, for the combination of lenses or lenses and mirrors one again needs to be cautious in the application of sign convention to image distance. For example when a 'real' image from the first component acts as a virtual object for the next component as was the case in the second example above, then the image distance for the first component is taken to be positive even though in reality the image from the first component was never formed, and even though the object distance for the next component using the same image is negative. A rule similar to the one stated above for object distance can be used to determine the sign of the image distance: *"If the image distance is measured in the same direction as the incident rays then image is real and the distance is positive. If the distance is measured opposite to the direction of the incident rays then the image is virtual and its distance is negative."*

Heights of the object (h_o) and the image (h_i) follow the Cartesian (rectangular) coordinates sign convention corresponding to the *y-axis, i.e.* height measured vertically upwards from the principal axis is positive $(+)$, and measured vertically downwards is negative $(-)$. Sign convention for magnification $m = h_i/h_o$ follows from the signs for heights. Magnification for an inverted image (which also is a real image as we saw from the ray diagrams) is negative, and for an upright (virtual) image *is positive*.

Finally, the sign convention for the radii of curvature of the lens surfaces is: *Looking at the lens along the direction of the incident ray, the radius of curvature is positive if the surface is convex, and it is negative if the surface is concave.* This sign convention can also be stated in the same way as the sign convention for the object distance, *i.e., if the radius of curvature is measured (from the center of curvature) opposite to the direction of the incident rays then it is positive, and if the radius of curvature is measured in the same direction as the incident rays then the radius of curvature is negative.* Figure 4.3 shows the signs of the radii of curvature of a bi-convex and a bi-concave lens when the rays are incident from the left. How would these signs be affected if the rays were incident from the right?

4.3.2 Refraction at a Convex surface

Consider two media of refractive indices n_1 and $n_2(> n_1)$, separated by a spherical surface of radius of curvature R, convex towards the rare medium (Figure 4.9). A ray of light from a point object *O* on the principal axis in medium 1, on refraction at point *P* at the interface bends towards the normal, and intersects the principal axis at point *I*, which is the real image of the object. Let x_o and

x_i be the distances of the object and the image measured from point S on the surface, and $d = PQ$ be the vertical distance of the point of incidence from the principal axis. Let i and r be the angles of incidence and refraction respectively, and α, β and γ are the angles of the incident ray, refracted ray, and the normal at point P respectively from the principal axis. We consider the point of incidence to be close to the principal axes, so that the points S and Q are very close together, and all the angles i, r, α, β, and γ may be treated as small angles.

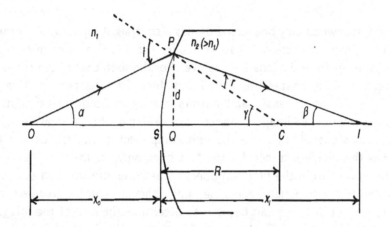

Figure 4.9: Refraction from a rare medium to a denser medium at a convex interface.

From Snell's Law, $n_i \sin i = n_2 \sin r$, and using $\sin i \approx i$ and $\sin r \approx r$ for small angles in radians:

$$n_1 i = n_2 r \tag{4.1}$$

From Figure (4.9):

$$i = \alpha + \gamma, \qquad r = \gamma - \beta \tag{4.2}$$

Substituting equation 4.2 in equation (4.1) and simplifying:

$$n_1 \alpha + n_2 \beta = (n_2 - n_1)\gamma \tag{4.3}$$

Now for small angles, α, β and γ:

$$\alpha \approx \tan \alpha \approx \frac{d}{x_o}, \quad \beta \approx \tan \beta \approx \frac{d}{x_i}, \quad and \quad \gamma \approx \tan \gamma \approx \frac{d}{R} \tag{4.4}$$

Substituting equation (4.4) in equation (4.3) and after simplifying we get:

$$\frac{n_1}{x_o} + \frac{n_2}{x_i} = \frac{n_2 - n_1}{R} \tag{4.5}$$

This is the equation of refraction at a convex surface separating two media, and relates the object and image distances to the radius of curvature of the interface and the refractive indices of the two media. We note that for a fixed object distance x_o, the image distance x_i is independent of the angle of incidence, so long as the angles are small. Thus all paraxial rays (rays close to the principal axis) come to focus at the same point. Equation (4.5) which is derived here for $n_1 < n_2$ can be applied equally well to the case $(n_1 > n_2)$, and x_i would turn out to be negative. Why? It is left as an exercise for students to derive the equation for the case $(n_1 > n_2)$.

4.3.3 Refraction at a Concave Surface

Figure 4. 10 shows the refraction of light from a rare medium (n_1) in to a denser medium (n_2) through a concave interface (on the rare medium side) of radius of curvature R. I is the virtual image of the object O.

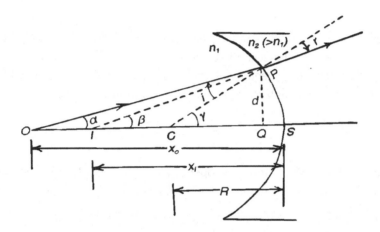

Figure 4.10: Refraction from a rare medium to a denser medium at a concave interface.

Applying the same assumptions, notations and principles as in subsection 4.3.2, we get:

$$i = \gamma - \alpha, \qquad r = \gamma - \beta \tag{4.6}$$

Substituting equation (4.6) in equation (4.1) and simplifying:

$$n_2\beta - n_1\alpha = (n_2 - n_1)\gamma \tag{4.7}$$

Now for small angles, α, β and γ:

$$\alpha \approx \tan\alpha \approx \frac{d}{x_o}, \quad \beta \approx \tan\beta \approx \frac{d}{x_i}, \quad and \quad \gamma \approx \tan\gamma \approx \frac{d}{R} \tag{4.8}$$

Substitute equation (4.8) in equation (4.7), and apply the sign convention, *i.e.* the image is virtual so that x_i is replaced with $(-)x_i$, and the radius of curvature R for a concave surface seen from the incidence side is negative. On simplifying we get:

$$\frac{n_1}{x_o} + \frac{n_2}{x_i} = \frac{n_2 - n_1}{R} \tag{4.9}$$

This equation is the same as equation (4.5) and a similar discussion applies. Students are once again assigned to draw the ray diagram and derive this equation for $(n_1 > n_2)$ case for the set up of Figure 4.10.

4.3.4   Lens Maker's Formula

Lens maker's formula gives the power of a lens (inverse of the focal length) in terms of the refractive indices of the material of the lens *(n)* and the medium (n_o) in which it is to be used, and the radii of curvature of the two surfaces of the lens. Consider a double convex lens with r_1 as the radius of curvature of the incidence-surface of the lens, and r_2 as the radius of curvature of the surface from which the refracted ray emerges. The required formula is obtained by combining equations (4.5) and (4.9) for refraction at the two surfaces of the lens.

For refraction at the incident surface, if we denote the image distance as x_i', then from equation (4.5):

$$\frac{n_o}{x_o} + \frac{n}{x_i'} = \frac{n - n_o}{r_1} \quad , \tag{4.10}$$

For refraction at the second surface, the image formed from the first surface acts as the virtual object for the second surface, *i.e.* for the second surface $x_o = (-)x_i'$. Furthermore, the object for this surface is effectively located in the medium of the lens, and the final image is formed in the outside medium, *i.e.* $n_1 = n, n_2 = n_o$. Substituting these in equation (4.9) we get:

$$(-)\frac{n}{x_i'} + \frac{n_o}{x_i} = \frac{n_o - n}{r_2} \tag{4.11}$$

Why and how? Students are assigned to draw a proper ray diagram for this exercise and to verify equations (4.10) and (4.11). Adding equations (4.10) and (4.11), and using the lens equation (derived later in section 4.4.1):

$$\frac{1}{f} = \frac{1}{x_o} + \frac{1}{x_i}, \tag{4.12}$$

one gets the lens makers formula:

$$\frac{1}{f} = \left(\frac{n}{n_o} - 1\right)\left(\frac{1}{r_1} - \frac{1}{r_2}\right) \tag{4.13}$$

The formula derives its name from the fact that it is used by a lens maker to grind a lens of required power from the material of known refractive index. If the radius of curvature of one of the surfaces is fixed (let us say that it is taken to be plane, or the radius of curvature of both the surfaces are taken to be equal etc.) the radius of curvature of the other surface can be determined from equation (4.13). The formula also assumes that medium on both sides of the lens is the same, which generally is the case. Although we have derived the lens makers formula for a double convex lens, it applies to any combination of surfaces namely convex, concave and plane provided correct sign convention is applied. Repeat the exercise to derive this formula for a double concave lens.

4.4   Analytical Location and Characterization of Image from a Lens

Although one can locate and characterize image formed by a lens by drawing a to-the-scale ray diagram, ray diagrams have geometrical limitations, and the accuracy of results depends on the accuracy in the use of scale for drawing and measuring. The problem becomes even more complex when two or more lenses or combinations of lenses and mirrors are involved, where image from one optical component acts as an object for the next imaging component. Therefore, analytical location and characterization of images formed by single or combinations of lenses or lenses and mirrors is indispensable. In this section we derive the lens equation (equation 4.12)which relates the distances of the object (x_o), and the image (x_i) and the focal length (f) of the lens. Further, magnification is expressed as the ratio of the image and object distances.

4.4.1   Derivation of Lens Equation

The lens equation can be derived by using any one of the seven ray diagrams given in section 4.2.1. Here we shall use the ray diagram for the divergent lens (Figure 4.7), which is a slightly difficult case because this case requires a careful application of the sign convention in the derivation of the equation. Other than that the actual mathematical procedure is rather straight forward that uses the properties of similar triangles. We reproduce Figure 4.7 as Figure 4.11 below for this exercise.

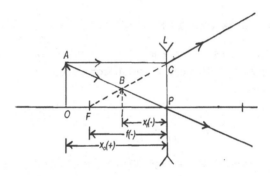

Figure 4.11: Ray diagram for the derivation of the lens equation.

From Figure 4.11: $\triangle$s OAP and IBP are similar, so that the ratio of their sides are equal, *i.e.*

$$\frac{IB}{OA} = \frac{PI}{PO},$$

$$or \quad \frac{h_i}{h_o} = \frac{-x_i}{x_o} \qquad (4.14)$$

Similarly, from similar $\triangle$s BIF and CPF,

$$\frac{IB}{PC} = \frac{IF}{PF} = \frac{PF - PI}{PF},$$

$$or \quad \frac{h_i}{h_o} = \frac{-f - (-x_i)}{-f} = \frac{f - x_i}{f} \qquad (4.15)$$

where in equations (4.14) and (4.15) we have used sign convention for distances x_i and f. From equations (4.14) and (4.15):

$$\frac{h_i}{h_o} = \frac{-x_i}{x_o} = \frac{f - x_i}{f},$$

$$or \quad x_o x_i = f x_i + f x_o \qquad (4.16)$$

dividing by fx_ox_i on both sides

$$\frac{1}{f} = \frac{1}{x_o} + \frac{1}{x_i} \qquad (4.17)$$

This is the lens equation which relates f, x_o and x_i. If any two of f, x_o and x_i in equation 4.17 are known, the third quantity can be calculated. When the sign convention for the given distances is used correctly, sign of the calculated quantity tells us whether it is real or virtual.

From equation (4.14) the magnification *(m)* in terms of image and objects distances is given as:

$$m = \frac{h_i}{h_o} = (-)\frac{x_i}{x_o} \qquad (4.18)$$

from which one can calculate the magnification and height of the image if the height of the object is known. The sign of magnification gives the characteristics of the image, *i.e.* whether the image is real or virtual and inverted or upright with respect to the corresponding object.

4.4.2   Focal length of a Thin lens Doublet

Consider a lens doublet of two thin lenses of focal lengths f_1 and f_2 placed in contact with each other. Thickness of the lenses is assumed to be very small compared to their focal lengths. Focal length*(F)* of the lens doublet is given by:

$$\frac{1}{F} = \frac{1}{f_1} + \frac{1}{f_2} \qquad (4.19)$$

Equation (4.19) can be derived by treating the image formed by the first lens as the object for the second lens, and by ignoring the thickness of lenses compared to the image and object distances and focal lengths. It is left as an exercise for students.

4.5   Defects of Lenses

So far we have directly or implicitly assumed that spherical lenses have a sharply defined focal point, and that the refractive index of the material of the lens is the same for all wavelengths of light. Three rays that were used to locate and characterize images geometrically were just approximations, though good ones for thin and small lenses but not so ideal for large and thick lenses. In practice any of these idealization and approximations do not apply. On the contrary, spherical lenses have a number of *aberrations* (imperfections) which impart defects to the image, and the image is far from the theoretically presumed ideal, perfect image. The defects become more prominent for large and thick lenses.

Lens aberrations can be grouped into two categories: *monochromatic aberration* and *chromatic aberration*. Monochromatic   aberrations are present even when a monochromatic light is used, and depend on the size, shape and curvature of the lens. Amongst these aberrations, the ones which we shall discuss here are the *spherical aberrations* and *astigmatism*. The chromatic aberration arises from the different refractive index of the material of the lens for different wavelengths, and comes in to play when polychromatic light such as white light is used.

4.5.1 Spherical Aberrations

When a monochromatic beam of light parallel to the principal axis is refracted through a lens, rays at different distance from the principal axis focus at different points on the principal axis. This is so because the assumptions made for a small lens, for example the small angles assumptions in subsections 4.3.2 and 4.3.3, do not apply for a lens of large aperture, and the refraction is solely governed by the fundamental laws of refraction. Consequently rays that are closer to the principal axis of the lens have a larger focal length than the rays that are farther from the principal axis. Thus the focal point of a large aperture lens is not sharply defined, it is spread along the principal axis (Figure 4.12), and the image of an object located at infinity is not a point image rather it is confined within a disk of finite size in between the two extreme focal points. This defect of lenses is known as the spherical aberrations. Defect in the image arising from spherical aberrations of lens can be minimized, but not completely eliminated by using lenses of small aperture (compared with the focal length of the lens). Smaller is the ratio of the aperture of the lens to its focal length, closer the lens shall be to being free from spherical aberration. Ideally, a pin-hole camera fitted with any lens shall be as close to being perfect as it can get, but this limits the amount of light passing through the lens, and the brightness of the image. Those who had the opportunity to use costly, professional photographic cameras with variable aperture would be familiar that a smaller aperture of the lens gives a sharper image with greater depth of field, but one requires longer exposures. In practice, a lens of aperture equal to one fifth of its focal lens is adequate for routine applications in microscopes, telescopes, fixed focus cameras, binoculars etc.

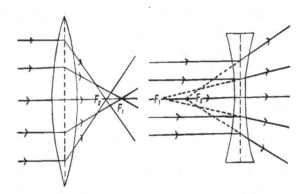

Figure 4.12: Spherical aberrations of bi-convex and bi-concave lenses.

4.5.2 Astigmatism

Astigmatism, also known as the off-axis astigmatism, occurs for objects not located on the principal axis of the lens. An off axis point object *(O)* is imaged as two different lines *HH'* and *VV'* in the horizontal and the vertical planes respectively, known as the focal lines (Figure 4.13). The final image viewed is a circle of least confusion lying between the two extreme focal lines in the horizontal and vertical planes in which neither the vertical nor horizontal details of the object are in sharp focus. When astigmatism as a defect of vision is present, the eye is unable to focus horizontal and vertical lines of a rectangular grid simultaneously, *i.e.*, if the horizontal details of the view are in focus, the vertical details are blurred, and vice versa. This defect of vision is corrected by the use of cylindrical lenses.

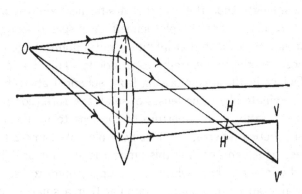

Figure 4.13: Astigmatism of a lens.

The other monochromatic aberrations of lenses not discussed here are: *curvature of field* which results because different points of a flat object are not equidistant from the lens, *distortion* occurs due to different magnification for points located at different distances from the lens axis, and could result either in *barrel distortion* or *pincushion distortion*, and *coma* when the image of an off-axis point objects results in a comet like image rather than a circle.

4.5.3 Chromatic Aberrations

This defect of lenses arises from different refractive index of the material of lens for different wavelengths, and comes into play when polychromatic light such as white light is used. The shorter wavelengths at the blue end of spectrum have a shorter focal length, and the longer wavelengths at the red end of spectrum have longer focal length. The intermediate wavelengths are focused between the two extreme focal points for the blue and red lights.(Figure 4.14) The result is an image spread along the principal axis with hues of different colours. The chromatic aberration for any two wavelengths is eliminated by using two lenses cemented together, normally one converging and the other diverging, made of different materials. The different refractive indices for the materials of lenses are such that they compensate favorably for the wavelength dependence of focal length. Powers of the convergent and the divergent lenses depend on the power required for the doublet. By choosing the wavelengths appropriately, the aberration can also be minimized for other wavelengths. Such a combination of two lenses is known as the *achromatic doublet*.

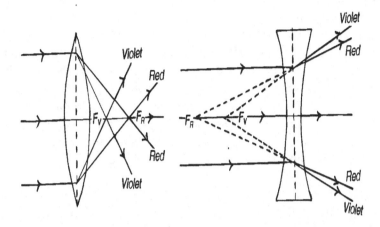

Figure 4.14: Chromatic aberrations of lenses.

4.6 Chapter 4 Summary

The Chapter deals with the properties of, refraction through, and image formation from lenses. Lenses are refractive medium bounded with two surfaces at least one of which is a curved surface. In this chapter we study the spherical lenses bounded by at least one spherical surface. The other surface may be plane or spherical as well. our focus shall be the (double) convex lenses with both surfaces convex, and the (double) concave lenses with both surfaces concave. They are also known as convergent and divergent lenses respectively. The lenses have a circular cross section, diameter of which is known as the aperture of the lens. We shall be dealing with thin lenses of small aperture (compared to the focal length).

- The centers of curvature, C_1 and C_2, and the radii of curvature, r_1 and r_2, refer to the corresponding centers and radii of the spherical surfaces of the lens. For a plane surface C lies at ∞, and the corresponding $r = \infty$. Line joining C to any point on the corresponding spherical surface of the lens is normal to the surface. Why?

- Line joining C_1 and C_2 through the lens is the principle axis of the lens.

- *Optical center* of the lens is a point P of the principle axis. A ray of light passing through P is refracted through a thin, small lens un-deviated, (to first approximation). For a symmetric lens it lies symmetrically inside the lens, but it may also lies just outside the lens. All distances along the principle axis, namely object and image distances, and the focal length are measured from P. However, the radius of curvature is measured from the surface of the lens.

- Refraction through a lens follows the usual laws of refraction as discussed in Chapter 2, but refraction of three rays is of particular interest. Any two of these rays can be used to locate and characterize the image by geometrical method from a ray diagram drawn to scale. Different scales are used for distances along the principal axis (x-axis) and lengths transverse to the principle axis (y-axis). A common mistake by students is to use the same scale for both.
  ○ Ray 1 incident parallel to the principle axis on refraction through a convex lens passes through (converges to) a point, F, on the principle axis on the other side of the lens. In case of the concave lens the incident ray appears to diverge from point F on the principle axis on the incidence side. F is known as the focal point of the lens; it is a real point for the convex lens and a virtual point for the concave lens. Distance PF is the focal length, f, of the lens. A lens has two focal points, F1 and F2 on both sides of the lens, and correspondingly there are two focal lengths, f_1 and f_2. When the medium on both side of the lens is the same, say air, then $f_1 = f_2 = f$.
  ○ Incident Ray 2 passing through P is refracted through both type of lenses un-deviated. Why?
  ○ Incident Ray 3 passing through F on the incident side of the convex lens, or incident in the direction of F on the other side of the concave lens is refracted parallel to the principle axis. This, together with Ray 1 illustrates the reversibility of the path of light.

- Power of a lens is defined as: $P(Diopter) = 1/f(m)$.

- Six cases of image formation from a convex lens are of interest: (i) $x_o = \infty$: Image is real, inverted, reduced in size and lies at distance $f(F)$ on the other side of the lens. (ii) $\infty > x_o > 2f$: Image is real, inverted, reduced in size and lies between distance $2f$ and $f(F)$. (iii) $x_0 = 2f$: Image is real, inverted, same size as the object and lies at distance $2f$. (iv) $2f > x_o > f$: Image is real, inverted, enlarged in size and lies beyond $2f$ distance. (v) $x_o = f$: Image is real, inverted, enlarged in size and lies at ∞. (vi) $x_o < f$: Image is virtual upright, enlarged in size and lies behind the object on the same side of the lens.

- For a concave lens there is only one case of image formation. For any location of the object, the image is virtual, upright, reduced in size and lies between the focal point and the lens.

- Refraction from a rare medium to a denser medium at a convex or a concave interface are both expressed mathematically as: $(n1/xo + n2/xi) = (n2n1)/R$.

- Lens makers formula gives the power of a lens and is used for grinding a lens of desired focal length. It is given as: $1/f = (n/no1)(1/r11/r2)$.

- Analytically one uses the lens equation; $(1/x_o) + (1/x_i) = (1/f)$, and the expression for magnification: $m = (h_i/h_o) = (-)(x_i/x_o)$ in conjunction with a sigh convention to locate and characterize the image.

- Focal length of a thin lens doublet (placed in contact with each other): $(1/F) = (1/f_1) + (1/f_2)$.

- All the formulae follow a Sign convention which simply stated is: All real distances are positive, and all virtual distances are negative.
  ○ f is (+) for a convex lens, and (-) for the concave lens.

○ x_o for a single lens is always (+), but for a combination of lenses / lenses and mirrors it could be (-) if the intermediate image is virtual.

○ x_i is (+) for cases (i) to (v) of the convex lens, and (-) for case (vi) of the convex lens and for the concave lens. For a combination of lenses / lenses and mirrors mirrors care must be taken to choose the correct sign.

○ m is (+) for an upright and virtual image, and (-) for an inverted and real image.

○ Sign of the radii of curvature r of the lens surface: *Looking at the lens along the direction of the incident ray, r is (+) if the surface is convex and (-) if the surface is concave.*

○ When the known quantities are used with a correct sign, the calculated quantity comes out with the correspondingly correct sign, which is a cross-check for the correctness of signs used in calculation.

- Defects of Lenses (Lens Aberrations): Following three are discussed briefly.

  ○ *Spherical Aberration:* Rays far from the principle axis are focused at a shorter distance than the rays closer to the principle axis. This results in a diffused focal point. The effect can be minimized by using a lens of smaller aperture compared to its focal length.

  ○ *Astigmatism:* An off axis point object is focused as different lines in the horizontal and vertical lines, resulting a blurred image circle. The defect is corrected by using cylindrical lenses.

  ○ *Chromatic Aberration* results from the wavelength dependent refractive index of the medium of the lens and hence the wavelength dependent focal length of the lens. The resulting white light image has streaks of colour spread over the range of focal length. This is corrected by using a achromatic doublet of two lenses.

4.7 Exercises

4.1 A diverging lens of focal length 10 cm is placed at a distance of 30 cm to the right of a converging lens of focal length 15 cm. A 5 cm high object is placed at a distance of 40 cm to the left of the converging lens. Locate and characterize the final image formed by the combination of lenses by drawing a to the scale ray diagram.

4.2 Repeat exercise 4.1 if the separation between the two lenses is 20 cm.

4.3 Repeat exercises 4.1 and 4.2 if (i) both the lenses are converging, (ii) both the lenses are diverging, (iii) lens on the left is diverging and the lens on the right is converging.

4.4 Calculate the focal length of the following lenses made from the material of refractive index 1.5, and used in air (take $n_o = 1$). (i) A double convex lens with radii of curvature of surfaces 20 cm and 30 cm, (ii) a double concave lens with radii of curvature of surfaces 20 cm and 30 cm, (iii) a plano-convex lens with radius of curvature of the spherical surface 20 cm, (iv) a plano-concave lens with radius of curvature 30 cm, (v) a plane window glass of thickness 0.6 cm.

4.5 Repeat the calculations of exercise 4.4 if the lenses are used under water, given $n_w = 1.33$.

4.6 A hemispherical glass block of radius of curvature 10 cm, and refractive index 1.47 is placed flatly on a printed page. Calculate the location of the image of the print on the paper from the top surface of the sphere for the print (i) at the center of the sphere, (ii) at 4 cm from the center, and (iii) at 8 cm from the center.

4.7 Parallel rays of light are incident on a glass sphere of radius 20 cm and refractive index 1.5. Consider two rays separated by 10 cm, and one of them passing through the center of the sphere. Calculate distance of the point from the center of the sphere where the two rays come to focus.

4.8 In exercise 4.7 consider two parallel rays, one through the center of the sphere, and the other incident at 10° at the surface of the sphere. Calculate distance of the point from the center of the sphere where the two rays come to focus. Compare the two focal lengths calculated in exercise 4.7 and 4.8, and discuss the features of the focal length of a thick lens.

4.9 A lens of power +4 D is held at a distance of 10 cm from a printed page. If the height of the printed line is 2 mm, what is the height of the print when viewed through the lens. Repeat the calculations for a lens of power $-4D$. Which of the two lenses can be used as a magnifier? Calculate its focal length, and name the lens in terms of the convergence/ divergence property of the lens.

4.10 Solve exercises 4.1, 4.2 and 4.3 analytically and compare your results to the results obtained by graphical method.

4.11 A 2 cm high object is placed 30 cm to the left of a convex lens of focal length 20 cm. A concave lens of focal length 30 cm is placed to the right of the first lens. Calculate the location and height of the final image formed by the combination of lenses, and characterize the image. Verify your results by plotting a to-the-scale ray diagram.

4.12 A 4 cm high object is placed at a distance of 30 cm to the left of a convergent lens of focal length 20 cm. To the right of the lens, at a distance of 100 cm a spherical mirror of focal length +30 cm is placed. Calculate the location, and size of the final image formed after one reflection from the mirror, and characterize the image.

4.13 Repeat exercise 4.12 if the separation between the lens and the mirror is 40 cm.

4.14 An object is placed at a certain distance from a convergent lens, and the image falls on a screen 60 cm from the lens. When a divergent lens is introduced halfway between the convergent lens and the screen, the screen must be moved 15 cm further away from the lenses to obtain a sharp focused image. Calculate the focal length of the divergent lens.

4.15 An image one third the height of the object is formed on a screen placed at a distance of 10 cm from a lens. Calculate the distance of the object and focal length of the lens. Keeping the lens fixed, by how much and in which direction should the object and the screen be moved to obtain an image three times the object height?

Chapter 5

Optical Instruments

In this chapter we discuss the applications of the laws of geometrical optics to optical instrumentation. The following optical instruments are discussed.

- Human eye

- Simple camera

- Simple microscope

- Compound microscope

- Keplerian telescope

- Galilean telescope

- Newtonian reflecting telescope

5.1 The human eye

The human eye is one of the marvellous gifts of nature. Large treatises and volumes have been written about the eye, covering several aspects of biology, optics, biophysics, photophysics, photochemistry, psychology and many other branches. In this chapter, a brief treatment of the optical features of the eye as well as common optical defects and their correction shall be considered.

5.1.1 Structure of the human eye

The basic structure of the eye is illustrated in Figure 5.1. The various parts are described briefly.

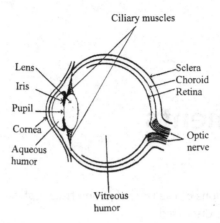

Figure 5.1: Basic structure of the human eye.

The cornea is a transparent membrane in the front surface of the eye through which light enters.

The iris is a circular diaphragm whose aperture is adjustable, and is sensitive to varying light intensities.

The pupil is the part of the eye which contracts in presence of a high intensity of light and dilates when there is low intensity.

The lens of the eye has a curvature or focal length that is adjustable by the spontaneous action of the ciliary muscle. The ability of the eye of the lens to have variable focal length is referred to as *accomodation*.

The ciliary muscle is the muscle that supports the lens in the eyeball.

Aqueous humor is a liquid in front of the eye lens.

Vitreous humor is a liquid at the back of the eye lens.

The retina is the sensitive sensor membrane consisting of cones and rods which receive light signals and convert them to electrical signals. This is analogous to the sensitive film in a camera. The cones are known to be sensitive to colour and bright light, while the rods are sensitive to faint light, motion and intensity variations.

The sclera is the thick white outer casing of the eye.

The choroid is a pigmented black membrane that absorbs stray light.

The Optic nerve is part of the nervous system and transmits the electrical signals generated in the eye to the brain.

For a healthy eye, the *far point* is practically infinity. If the eye is viewing a distant object, the muscles are relaxed, and the lens has a long focal length. On the other hand, if the eye is viewing a nearby object, the muscles are contracted, and the lens has a shorter focal length. The shortest attainable focal length determines the shortest distance for which the eye can see objects clearly. This shortest distance is known as the *near point*, which for a normal young adult is approximately 25 cm. With advancing age, the near point increases.

5.1.2   Common optical defects of vision and their correction

The common optical defects of the eye are far sightedness, short sightedness, astigmatism, coma, glaucoma, cataract, night blindness. The first two are discussed below, and illustrated in Figure 5.2.

Far sightedness (also known as *hypermetropia* or *hyperopia*) is a condition of the eye whereby distant objects are seen clearly but close objects are unclear. This can be understood as follows. Light from a near object striking the eye is out of focus and the image is formed not on the retina, but behind it, as illustrated in Figure 5.2 (a). This can be attributed to a small or flattened eyeball such that the focal length of the eye lens is long.

Short sightedness (also known as *myopia*) is a condition of the eye whereby near objects are seen clearly but distant objects are unclear. In this case, light from a far object striking the eye is out of focus and the image is formed not on the retina, but before it, as illustrated in Figure 5.2 (b). This can be attributed to a large or elongated eyeball such that the focal length of the eye lens is short.

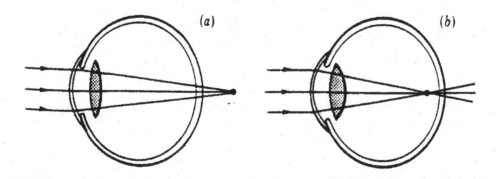

Figure 5.2: Common eye defects: (a) Far sightedness (*hypermetropia*), (b) Short sightedness (*myopia*).

The corrections to far sightedness and short sightedness are accomplished by using lenses, as illustrated in Figure 5.3. Far sightedness is corrected by using a convex lens (converging) and short sightedness is corrected by using a concave lens (diverging).

Focal length (power) of the corrective lenses can be calculated as follows, if the defective vision distances are known.

(a) *Power of the corrective lens for far sightedness*: Let d (¡ d_{near}, 25 cm for a healthy eye) is the near point of the defective eye. One uses a lens which produces a virtual image of an object placed at d at the d_{near} point. Thus from the lens formula: $(1/f) = (1/d)(1/d_{near})$ gives the power of the corrective lens, where all distances are in m. Since $d < d_{near}$, the lens is convergent.

(b) *Power of corrective lens for short sightedness*: Let $d(< \infty)$ is the far point of the defective eye. One uses a lens which produced the virtual image of a far object at ∞ at a distance d, the far point of the defective eye. Then, from the lens formula: $(1/f) = (1/\infty) + (1/-d) = (-1/d(m))$ gives the power of the corrective lens, which is a divergent lens.

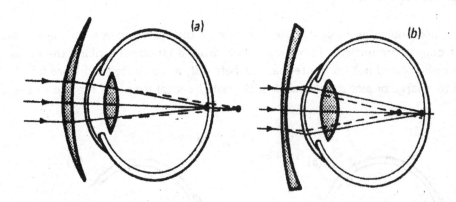

Figure 5.3: Eye defects corrections: (a) Far sightedness is corrected by using a convex lens, (b) Short sightedness is corrected by using a concave lens.

Astigmatism is the inability of the eye lens to focus simultaneously light rays arriving in different planes. This is caused by a slight distortion in the curvature of the cornea. Correction of this defect requires a lens with a cylindrical surface, which focuses the rays of light properly. Astigmatism, as a defect of lenses, was discussed in section 4.5.2.

5.2 A simple camera

The basic principle of a camera is that there must be an object, a convex lens and a film which captures a real image of the object. The basic structure of a camera is illustrated in Figure 5.4.

The working of certain parts of the eye and those of the camera are somewhat analogous, as illustrated in Table 5.1.

Table 5.1: A comparison between certain parts of the eye and those of the camera.

Eye parts	Camera parts
Eye lens	Glass lens
Pupil	Aperture
Iris	Diphragm
Retina	Film

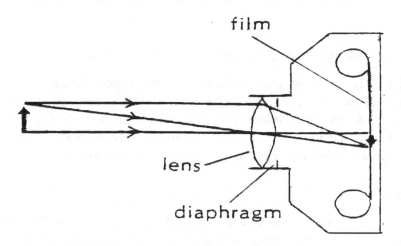

Figure 5.4: Basic structure of the camera.

The various parts of the camera are described below briefly.

The lens of the camera is a convex lens with a fixed focal length, but its position is adjustable, and practically the nearer the object, the greater the distance from the lens to the film. There is a certain minimum distance of the object from the lens for a clear image to be formed on the film. Why? The movement of the lens to form a clear, sharp image is known as *focusing*. Some simple cameras have a fixed focus. *Note that* the focusing of a camera is not the same as the focal length of the lens. High quality expensive cameras consist of a combination of lenses to eliminate the defects of the image due to lens aberrations, discussed in chapter 4.

In fact, image formation by a simple camera is analogous to case (ii), Figure (4.6 b), Chapter 4, of image formation by a convex lens whereby the object is placed at a distance $> 2f$ and the image is formed at a distance between f and $2f$ from the lens, where the photographic film is placed in the camera. This is achieved by focusing the camera lens, and since a camera lens has a small focal length (large power), only a small movement of the lens is needed to focus the image on the film.

The film is the chemical sensor which captures the image. In comparison to the eye, the film is analogous to the retina. In a digital camera, the film is replaced by a CCD (Charge Coupled Device).

The aperture is the opening in the camera that allows light in and is controlled by the diaphragm.

5.3   Microscopes

Microscopes are instruments which are used to produce an enlarged image of a small object. In this section, two types of optical microscopes are discussed: the simple microscope and the compound microscope.

5.3.1   Angular size of an object and image.

Angular size of an object/ image is the angle subtended by the object/ image on the eye or the lens. If h is the height of the object/ image and d is the distance, then its angular size is: θ (radians)$\approx \tan\theta = (h/d)$. As an object/image of the same height approaches the eye, its angular size increases, and if it is moved away from the eye the angular size decreases. This is why the parallel railway tracks appear to be converging as one sees them at a distance.

5.3.2   Simple microscope

The simple microscope consists of a single convex lens, as illustrated in Figure 5.5. The object is located just inside the focal point of the convex lens, and the image is virtual, enlarged and upright. The simple microscope is also known as a "Magnifier".

In the most efficient mode of the use of a magnifier, one hold the object from the lens at a distance $d(< f)$ such that the virtual upright image is formed at the near distance of clear vision (near point, 25 cm for a healthy eye), d_{near}. Then from the lens equation: $(1/f) = (1/d) + (-1/d_{near})$, the magnification is obtained as: $m = (-)(-d_{near}/d) = 1 + (d_{near}/f)(d_{near}/f)$. Typically, if $f = 5cm$, and $d_{near} = 25cm$ for a healthy eye, magnification of up to 6 can be achieved from a simple magnifier. Smaller focal length cannot be used because of the limitation posed by spherical aberration of a thick spherical lens.

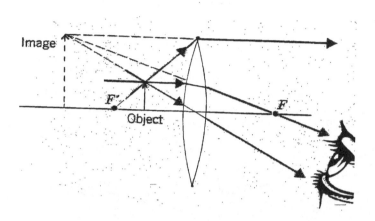

Figure 5.5: Principle of the simple microscope (magnifier).

5.3.3 Compound microscope

The compound microscope consists of two lenses, the one nearer the object is called the "objective", and another near the eye is known as the "eyepiece", as illustrated in Figure 5.6. Both of these lenses have a small focal length and are separated by an adjustable distance. The object is located just outside the focal point of the objective, and its image is formed just inside the focal length of the eyepiece.

As was mentioned in equation (3.1), there are two common notations for denoting object and image positions:

$$x_0 \leftrightarrow u$$
$$x_i \leftrightarrow v$$

In this chapter, for convenience, we shall use the uv notation. For the objective, the following parameters are applicable: u_0 denotes the object distance from the objective, v_0 denotes the image distance from the objective, and f_0 is the focal length of the objective. These parameters are related using the lens equation,

$$\frac{1}{u_0} + \frac{1}{v_0} = \frac{1}{f_0}$$

where $u_0 \geq f_0$. Magnification by the objective, $m_0 = -v_0/u_0$

Let the distance between the objective and eyepiece be D. For the eyepiece, the following parameters are applicable: u_e denotes the intermediate image distance from the eyepiece, v_e denotes the final image distance from the eyepiece, and f_e is the focal length of the eyepiece. These parameters are related using the lens equation,

$$\frac{1}{u_e} + \frac{1}{v_e} = \frac{1}{f_e}$$

where $u_e = (D - v_0) < f_e$.

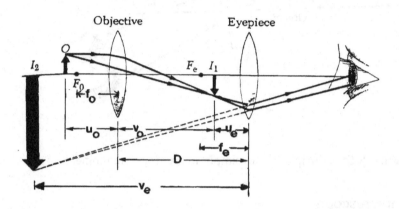

Figure 5.6: Basic structure of a compound microscope.

Angular Magnification by the eyepiece, $m_e = Near\ point/u_e$, where the near point is approximately 25 cm, as discussed in section 5.1.1.

Total angular magnification, M, is given by

$$
\begin{aligned}
M &= m_0 m_e \\
&= -\left(\frac{v_0}{u_0}\right)\left(\frac{Near\ point}{u_e}\right)
\end{aligned}
\tag{5.1}
$$

and using the approximations $u_0 \approx f_0$, $u_e \approx f_e$ and $v_0 \approx (D - f_e)$, the total angular magnification of the compound microscope can be written in the form

$$M \approx -\left(\frac{D - f_e}{f_0}\right)\left(\frac{Near\ point}{f_e}\right) \approx -\left(\frac{D}{f_0}\right)\left(\frac{Near\ point}{f_e}\right) \tag{5.2}$$

5.4   Telescopes

Telescopes are instruments used to view distant objects. In this section, we discuss the two main types of telescopes: the refracting telescopes and the reflecting telescope.

5.4.1 The Keplerian Telecope (a Refracting telescope)

The Keplerian telecope , also known as the astronomical telescope, consists of two convex lenses, the first one facing the object, known as the "objective", and the other one near the eye is known as the "eyepiece", as illustrated in Figure 5.7. The objective has a large focal length, f_0, and large aperture to receive maximum light, whereas the eyepiece has a small focal length, f_e. The distance between the objective and eyepiece is approximately $f_0 + f_e$.

The object is at a distant position from the objective. The objective forms a real, inverted intermediate image at the focal point of the objective, and this intermediate image is formed just inside the focal length of the eyepiece. The final image is large and inverted, and ideally at the least distance of clear vision, $d_{near} \approx 25$ cm.

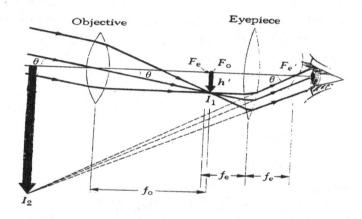

Figure 5.7: Basic structure of the Keplerian telecope (a refracting telescope).

For the objective, the following parameters are applicable: v_0 denotes the image distance from the objective, and f_0 is the focal length of the objective. These parameters are related using the lens equation,

$$\frac{1}{u_0} + \frac{1}{v_0} = \frac{1}{f_0}$$

where $u_0 \to \infty$, and hence $v_0 \approx f_0$.

For the eyepiece, the following parameters are applicable: u_e denotes the intermediate image distance from the eyepiece, v_e denotes the final image distance from the eyepiece, and f_e is the focal length of the eyepiece. These parameters are related using the following approximations: $u_e \approx f_e$.

Total angular magnification of the telescope, M, is given by

$$M = \frac{\theta_i}{\theta_0}$$

$$\approx \frac{h'/f_e}{h'/f_0}$$

$$\approx \frac{f_0}{f_e} \tag{5.3}$$

where for small angles the following approximation: $\theta_i \approx h'/f_e$ and $\theta_0 \approx h'/f_0$ can be used, as can be seen from the geometry of Figure 5.7.

5.4.2   The Galilean telescope (a Refracting telescope)

The Galilean telecope, also known as the opera glass, is another example of a refracting telescope and consists of an "objective" which is convex lens, and an "eyepiece" which is a concave lens, as illustrated in Figure 5.8. The objective forms an intermediate image I of the distant object and final image by the eyepiece is virtual, large and upright. The separation between the lenses is equal to the difference between the focal lengths of the objective and eyepiece, and the magnification is given by equation 5.3. The total length of the telescope is $(f_0 - f_e)$. This telescope is suited for viewing sports, distant wildlife, opera from a distant seat.

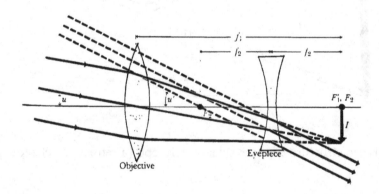

Figure 5.8: Basic structure of the Galilean telescope (a Refracting telescope).

5.4.3   The Newtonian Reflecting Telescope

The Newtonian Reflecting telecope consists of a parabolic mirror, plane mirror and an "eyepiece" which is convex lens, as illustrated in Figure 5.9.

Parallel rays from a distant object are incident on a parabolic mirror which reflects the rays towards a focal point. However, before the rays are brought to a focal point, they are intercepted by a plane mirror located within the telescope tube, and are then reflected through the eyepiece to an observer or some other detecting system. The final image is upside down and this telescope is suitable for astronomical applications. These telescopes can be made of very large aperture, sometimes in meters of aperture so as to collect more light from a distant object so as to produce a brighter image. The telescopes using lenses can not be made of very large apertures. Why?

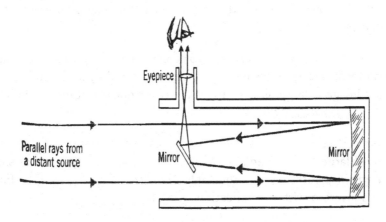

Figure 5.9: Basic structure of the Newtonian Reflecting Telescope.

5.5 Chapter 5 Summary

As applications of optical components and geometrical optics, the Chapter presents some common optical instruments. Construction, principle of working and performance indicators of the following instruments are discussed.

- Human eye is the most fascinating optical instruments of biological origin which all living being are endowed with. From optical perspective, eye comprises a biological convex lens of adjustable focal length and retina, the screen, on to which image of the observed object is projected. Because of illness or other reasons, some of the capabilities of eye may be diminished. Three defects of the eye can be conveniently corrected by external use of lenses of glass/ plastic worn in front of the eye. These defects are:
  ○ Far sightedness or hypermetropia: The near vision distance is reduced to d. A convex lens is used to correct this defect the power of which is given by: $P = (1/d)(1/d_{near})$.
  ○ Shortsightedness or myopia: Far objects are clearly visible only to a reduced distance d. Power of the corrective concave lens is: $P = (-1/d)$.
  ○ Astigmatism: This defect is correctable by use of cylindrical lenses.

- Photographic Camera: Its essential optical components are a convex lens for image formation, a photographic film for capturing the image, and a shutter to allow rays from the object to reach the film (screen). The film is located at a distance between f and $2f$ from the lens, the distance of which can be adjusted by moving the lens back or forward (focusing), and the object should be placed at a distance ¿ $2f$ for best results.

- Definition: Angular size of an object/ image is the angle (radians) subtended by it on the eye/ lens. Larger is the distance smaller is the angular size.

- Simple microscope, the magnifier is a convex lens of small focal length. object is placed at ¡ f distance from the lens and an enlarged, upright, virtual image is observed at dnear. The magnification obtainable is $\approx (d_{near}/f)$.

- Compound microscope: It is used to observe micro objects, and comprises an objective lens and an eyepiece of small focal lengths and small apertures, separated by a adjustable distance $D \approx (f_o + f_e)$. The final image is large, virtual and inverted at dnear distance. The magnification obtainable is: $\approx (-)(d/f_o)(d_{near}/f_e)$.

- Keplerian (astronomical) telescope is used to observe distant astronomical objects. It is made of an objective lens of large aperture and large focal length, and an eyepiece of small aperture and small focal length separated by an adjustable distance: $D \approx (f_o + f_e)$. Final image is enlarged, virtual and inverted with magnification: $m \approx ((f_o/f_e)$.

- Galilean Telescope or the opera Glass is used to observe theatrical events, wild life and sports from a distant seat/ location. The final image must obviously be enlarged and upright, but is virtual. The objective is a convex lens of a relatively large focal length and large aperture while the eyepiece is a concave lens of smaller focal length separated by adjustable distance $D \approx (f_o - f_e)$. The magnification obtained is $m \approx (f_o/f_e)$.

- The Newtonian Telescope: This is a reflecting telescope of up to meters wide aperture for receiving maximum light from distant astronomical objects for a brighter image. objective is a large parabolic mirror of relatively short focus, and eyepiece is a convex lens of shorter focus. A plane mirror is used to direct the image from the mirror to the eyepiece. Image is of course inverted (immaterial for astronomical objects), virtual and very large.

5.6 Exercises

5.1. A lens of focal length 80 mm is used to focus an image on the film of a camera. The maximum distance allowed between the lens and the camera is 12 cm.
(a) If the object to be photographed is 10.0 m away, calculate the distance between the lens and the camera for a sharp image on the film.
(b) How far from the lens is the closest object that could be photographed and produce a sharp image?

5.2 The lens of a 35 mm camera has a focal length of 55 mm. The distance of the lens from the film is adjustable over a range of 55 mm to 62 mm. Over what range of object distances as measured from the lens is the camera capable of producing sharp images?

5.3. The objective and eyepiece of a compound microscope have focal lengths 0.65 cm and 1.6 cm respectively. The tube length is 17.5 cm. Calculate the magnification of the compound microscope, assuming a normal near point.

5.4. The objective and eyepiece of a compound microscope have focal lengths 1.0 cm and 2.0 cm respectively. An organism to be examined microscopically is located at a distance of 1.1 cm from the objective. Calculate the length of the microscope tube and the magnification of the system, assuming a normal near point.

5.5. The objective and eyepiece of a compound microscope have focal lengths 0.2 cm and 2.5 cm respectively. The distance between the lenses is 18.0 cm. Calculate the angular magnification of the compound microscope, assuming a normal near point.

5.6. A Keplerian telescope has its two lenses spaced by 76.0 cm. If the objective lens has a focal length of 74.5 cm, what is the magnification of this telescope?

5.7. A Galilean telescope has its two lenses spaced by 33.0 cm. If the objective lens has a focal length of 36.0 cm, what is the magnification of this telescope?

5.8. The near point of a myopic eye is 40 cm. Calculate the power of the corrective lens to restore the near vision to normal.

5.9. The far point of a hyperopic eye is 2 m. Calculate the power of the corrective lens to restore the far vision to normal.

5.10. A simple microscope is held at a distance of 10 cm from a printed page, and the letters on the page appear to be 3 times the height of the print. Calculate the focal length of the magnifier.

5.11. A magnifier of focal length 5 cm is used by a person with a near point of 40 cm. Calculate the best position of the object when the magnifier is placed very close to the eye. What is the magnification achieved?

5.12. In an astronomical telescope the power of the objective lens is +2.0 diopters, and power of the eye piece is +75 diopters. Calculate the length of the telescope tube and its magnification.

5.13. In order to obtain the magnification of 75, what should be the power of the objective lens if the power of the eye piece is +40 diopters? What is the length of the telescope?

5.14. Which of the following combinations of lenses can be used for a microscope and which ones for the telescope?

Combination	1	2	3	4	5
f_o (cm)	1	10	0.3	20	10
f_i (cm)	2	2	1	4	-3

5.15. The objective of a large astronomical telescope has a focal length of 2.0 m. What should be the focal length of the eye piece to obtain a magnification of 100, 200, and 500?

Chapter 6

Interference of Light

Principles of geometrical optics and their applications discussed in Chapters 2 to 5 did not make any assumptions about the nature of light. In Chapter 2 we did, however, use Huygen's wave theory to derive the laws of refraction, and in classical electrodynamics, one derives these principles of optics from the electromagnetic-wave nature of light, and a brief introduction to electromagnetic theory was presented in section 1.4.2, in chapter 1. Nevertheless, the entire field of geometrical optics is fully understood and applied in terms of 'ray optics'. The ray represents the rectilinear propagation of light, and it applies equally well to either nature of light whether it comprises waves or photons. In contrast to this there are phenomenon exhibited by light which can not be explained by representing light as a ray along the direction of propagation. These phenomenon are the interference, diffraction and polarization, that constitute the *Physical Optics*, and explicitly require the wave nature of light for their understanding, or in the case of interference of subatomic particles such as electrons, then the wave nature through the de Broglie's hypothesis is invoked.

According to classical electrodynamics, light consists of transverse electromagnetic waves, *i.e.* oscillating coupled electric and magnetic fields propagating through the space with the speed of light. Both the fields are perpendicular to the direction of propagation, and are perpendicular to each other, as illustrated in Figure 1.7. Their amplitudes are linearly related as $B_o = \sqrt{\mu\epsilon}\,E_o$ where μ and ϵ are the permeability (magnetic) and permittivity (electric) of the medium respectively. Thus in terms of the Cartesian coordinates system, if $+x$ is the direction of propagation of light and the electric field vector is directed along the y axis then the magnetic field vector is directed along the z axis, *i.e.*, $\hat{k}\,B_o = \sqrt{\mu\epsilon}\,\hat{i} \times \hat{j}\,E_o$, where $\hat{i}$, $\hat{j}$ and $\hat{k}$ are the unit Cartesian vectors. Interference and diffraction are the direct evidence of the wave nature of light, whereas the transverse nature is established by polarization.

6.1 Superposition of Waves and Interference of Mechanical Waves

When two or more waves travel through a medium simultaneously, the resultant 'displacement of the medium' is the vector sum of the displacements due to individual waves. This is known as the *principle of the linear superposition* of waves and leads to interference of waves. Consider two waves traveling through a medium simultaneously. If at a given location and time maximum displacements due to individual waves coincide, the resultant displacement of the medium is equal to the sum of the two displacements. This is known as the *constructive interference*. On the other hand if the maximum displacement due to one wave coincides with the minimum displacement due to the other wave, the resultant displacement is equal to the difference of the two displacements in the direction of the larger displacement, and this is known as the *destructive interference*. This phenomenon can be seen quite clearly by throwing two small pebbles close to each other in a pool of water to generate two waves traveling through water simultaneously. Where the peaks or troughs of the two waves coincide, resultant displacement of the water surface is large, equal to the sum of individual displacements, and where peak due to one wave coincides with the trough due the other wave, the displacement of water surface is small, zero if the two displacements are equal in magnitude. In sound the interference manifests itself as the production of beats when two sound waves of frequency very close to each other travel simultaneously through the medium.

6.2   Interference of Light

In the case of light, interference can not be observed simply by placing two light sources side-by-side. Rather two beams of light must satisfy some stringent requirements for them to interfere constructively and destructively to produce a stationary interference pattern, and because of the very small wavelength of light ($\sim 10^{-7} m$), the maxima and minima of intensity are so close together that one invariably needs to use a microscope/ telescope to be able to see the interference pattern. Thus interference of light is observed only under controlled experimental conditions in a laboratory using specialized equipment. This is so because of a number of other reasons as well, amongst which the noteworthy ones are:

- Light from two sources is emitted independent of each other, and the two beams do not maintain a constant phase relationship. With the changing phase relationship, conditions of constructive and destructive interference at a given point of observation change so rapidly that no stable interference is obtained. Stated differently, the phase of light waves from an ordinary source change randomly about every $10^{-8}s$, and the interference pattern produced by light from such a source has a lifetime of only about $10^{-8}s$ which a human eye is unable to follow. The result seen by the eye is a uniform distribution of intensity.

- Amplitude of waves from two sources are also different so that destructive interference is only partial. If it was at all possible to see the interference pattern, the contrast between the points of maximum and minimum intensity shall be poor.

- Light from an ordinary source consists of a spectrum of wavelengths. Each wavelength produces an interference of its own, shifted with respect to each other, to give a more or less uniform distribution of intensity.

Such a pair of light sources is known as *incoherent*.

6.2.1   Coherent Source

In order to observe a stable interference pattern the pair of light sources must satisfy two conditions:

(i) The light sources must be *coherent*, *i.e.* the two interfering rays of light must maintain a constant phase relationship.

(ii) The light must be *monochromatic*, *i.e.*, of single wavelength so that only one interference pattern is obtained in the plane of observation which shall not be blurred from the overlap of interference patterns from different wavelengths.

In a later section we shall discuss interference using white light. Only a few low-order coloured interferences fringes due to various colours are seen, and those too under specific circumstances. The coherent sources are derived from the same source so that any random change or fluctuation in the original source is carried forward to the pair of coherent sources. In this manner the phase relationship between them always stays constant. This is achieved by any one of the following two methods.

Coherent sources by division of wavefront

From Huygen's wave theory (Chapter 2), light waves propagate through a medium as wavefronts, wavefronts are normal to the rays, and the wavefronts are the surfaces of same phase, *i.e.*, all points on a wavefront oscillate in phase. Furthermore, successive wavefronts from a point source (diverging rays of light) are concentric-spherical, and corresponding to a parallel beam of light they are plane parallel. To produce coherent sources, wavefront from a single source is divided laterally into segments either by using pair of slits as in the case of the Young's double slit interference experiment, or by refraction of a wavefront as in the case of Fresnel's biprism, or by reflection of a part of the wavefront from a mirror and other remaining part reaching the plane of interference directly from the source as in Lloyd's mirror experiment, or by reflection of wavefront from two plane mirror slightly inclined to each other as in the case of Fresnel's mirror. A point source or an illuminated fine slit is used so that the original wavefronts are either spherical or cylindrical. The intensity of both the interfering waves (power per unit area of the wavefront) can be set to be equal, and since the partitioned wavefronts are derived from the same wavefront, the phase relationship between them remains constant at all times, producing a stable interference pattern in the plane of observation.

Coherent sources by division of amplitude

The incident wavefront from the source is divided by partial reflection and refraction. Width of the resulting wavefronts remains the same as that of the incident wavefront but the intensity which is proportional to the square of the wave-amplitude is divided such that the incident intensity equals the sum of the intensities of the reflected and refracted components. Example of this are interference from thin films, Newton's rings, and the Michelson interferometer.

6.2.2 Monochromatic Sources of Light and the Lasers

The most commonly used monochromatic source of light to illustrate interference in laboratory is a sodium lamp. Although sodium light consists of two wavelengths in the yellow region of spectrum, the wavelengths are so close together (588.995 *nm* and 589.592 *nm*, $\Delta\lambda = 0.597$ *nm*) that one is able to see distinct interference fringes up to a high order from the stronger component of 588.995 *nm* wavelength. Observations from the overlap of interference patterns from two wavelengths is employed to determine the wavelength difference of the two components. Other sources of monochromatic light are the strong spectral lines from the line spectra of elements, for example mercury, helium, cadmium, hydrogen etc., or specific wavelengths obtained from polychromatic light such as white light by using high quality monochromating filters.

Lasers (Light Amplification by Simulated Emission of Radiation), discussed in Chapter 9, are not only highly pure monochromatic radiation, they are coherent source of light, and can be produced as a sharply focused, concentrated, narrow beam of light which can travel long distances with negligible spread. For example, a laser beam of 1 cm diameter produced on earth, on reaching moon 384,000 km away is no more than 500 m wide. Helium-neon lasers (λ =632.8 nm) are the commonly used lasers in teaching laboratories for interference experiments. Normally one uses a single laser beam to obtain two interfering beams either by division of wavefront as in the double slit experiment or by division of amplitude as with the Michelson interferometer. However, the most important and powerful aspect of the use of lasers in interference experiments arises from the fact that lasers are coherent source of monochromatic light. Laser beams from two separate sources maintain a constant phase relationship. Therefore, two separate lasers sources can be used in interference studies.

6.3   Phasor addition of waves

The resultant wave from the superposition of two or more waves traveling in a medium simultaneously is obtained by the vector addition of individual waves either analytically (algebraically) or graphically known as phasor addition of waves. Both methods have their advantages and disadvantages. Algebraic method gives the precise mathematical results, but it cannot always be applied conveniently if a large number of waves are involved. On the other hand graphical method is cumbersome in as far as drawing of to-the-scale phasor diagrams, and measurements of the resultant wave amplitude and phase angle are concerned, but one can add any number of waves in this manner without any mathematical difficulty. Once the amplitude and phase of resultant wave is known, the intensity at the point of observation is proportional to the square of the amplitude, from which conditions of maximum and minimum of intensities in the interference pattern can be deduced. We first deal with the phasor (graphical) addition of waves.

Light waves are electromagnetic in nature, and the oscillating electric field vector E can be expressed as:

$$E(x,t) = E_o \, sin \, (kx - \omega t + \alpha)$$

which is a *plane wave* propagating along $+x$ axis in the medium, where x is location, t is time, E_o is the amplitude of the wave, $k = (2\pi/\lambda)$ is the wavenumber, λ is the wavelength, $\omega = 2\pi\nu$ is the angular frequency, ν is the frequency and α is the initial phase. The speed of the propagation of waves in the medium is $v = \nu\lambda = \omega/k$. The *intensity* of light is given by the square of the amplitude of the electric field vector. The magnetic field vector B can always be obtained from the electric field vector.

The electric field vector at time t at a fixed location x along the direction of propagation can also be expressed as a simple harmonic motion: $E(t) = E_o \sin(\omega t)$ where we have also assumed the initial phase α of the wave to be zero. This wave motion is graphically represented by a phasor vector of magnitude E_o rotating anticlockwise in the y-z- plane with angular velocity ω (Figure 6.1 a), known as the phasor diagram of the wave motion. The electric field intensity at time t is given by the projection of E_o on the $y-$ axis. Likewise, if we have two wave motions, $E_1 = E_{o1}sin(\omega t)$ and $E_2 = E_{o2}sin(\omega t + \phi)$, their phasor diagram at time t is shown in Figure 6.1 b.

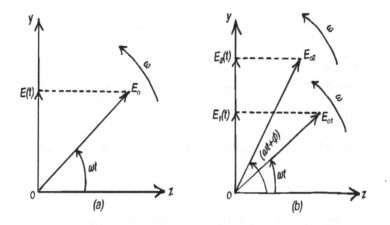

Figure 6.1: (a) Phasor diagram of a wave motion of amplitude E_o, and (b) phasor diagram of two waves with a phase difference of ϕ.

Now consider two wave motions with the same amplitude E_o and frequency ω arriving at the same point in space with a phase difference of ϕ (equation 6.1 and 6.2). The wave motion resulting from the superposition of these two waves is obtained by the phasor addition of the two waves as shown in Figure 6.2 (a). Considering both the wave motions at $t = 0$, wave 1 is represented with a phasor vector E_o along the horizontal, and the phasor vector of wave 2 is drawn at an angle ϕ from horizontal at the tip of the first phasor vector. The phasor vector of the resultant wave, E_{oR} and its phase angle is obtained by joining the tail of the first phasor to the tip of the second phasor. Students will recall that this vector superposition of two wave motions using phasors is exactly the same as the graphical addition of vectors.

$$E_1 = E_o sin(\omega t) \tag{6.1}$$

and

$$E_2 = E_o sin(\omega t + \phi) \tag{6.2}$$

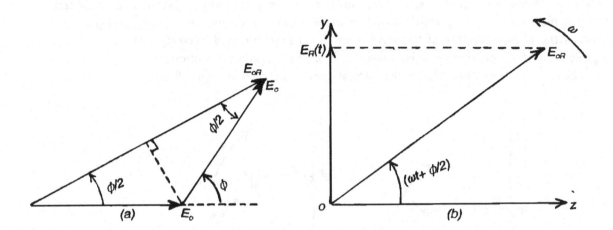

Figure 6.2: (a) Phasor addition of two wave motions of equal amplitude, (b) phasor diagram of the resultant wave.

Applying the principles of geometry one readily obtains the amplitude E_{oR} and the phase angle α of the resultant wave motions from Figure 6.2 (a) as:

$$E_{oR} = 2E_o cos(\alpha), \quad \text{and} \quad \alpha = \phi/2, \tag{6.3}$$

and the resultant wave motion is given by:

$$E_R = E_{oR} sin(\omega t + \alpha) = \left\{ 2E_o cos\left(\frac{\phi}{2}\right)\right\} sin\left(\omega t + \frac{\phi}{2}\right) \tag{6.4}$$

The phasor diagram of the resultant wave is shown in Figure 6.2 (b)

If we have N number of wave motions of amplitudes $E_{o1}, E_{o2}, E_{o3}...$ with initial phases $\phi_1, \phi_2, \phi_3...$, given as:

$$\begin{aligned}
E_1 &= E_{o1} sin(\omega t + \phi_1) \\
E_2 &= E_{o2} sin(\omega t + \phi_2) \\
E_3 &= E_{o3} sin(\omega t + \phi_3) \\
&\vdots \\
E_n &= E_{on} sin(\omega t + \phi_n) \\
&\vdots
\end{aligned} \tag{6.5}$$

their resultant can be found following the same procedure. Phasor addition of these waves, and the phasor vector E_{oR} and initial phase ϕ_R of the resultant wave are shown in Figure 6.3. These can be measured from the graph, and the resultant wave is given as:

$$E_R = E_{oR}sin(\omega t + \phi_R) \tag{6.6}$$

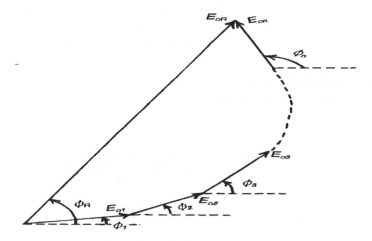

Figure 6.3: Phasor addition of N wave motions.

It is to be noted that phasor addition can be applied only to the waves of same frequency. Why?

6.4   Conditions of Interference

Intensity of the interference pattern produced from the superposition of waves is proportional to the square of the amplitude of the resultant wave, *i.e.* $I \propto E_{oR}^2$. Considering the superposition of two waves only (equation 6.1, 6.2), the intensity at the point of interference from equation 6.4 shall be:

$$I \propto 4E_o^2 cos^2\left(\frac{\phi}{2}\right),$$
$$\text{or} \quad I = I_{max}cos^2\left(\frac{\phi}{2}\right) \tag{6.7}$$

From equation 6.7, the intensity at the point of interference is maximum, ie $I = I_{max}$, if $\phi = 0, 2\pi, \ldots, n \times 2\pi$, and the intensity is minimum, equal to zero if $\phi = \pi, 3\pi, \ldots, (2n+1) \times \pi$ where $n(= 0, 1, 2, 3, ...)$ is an integer. Thus the conditions of constructive and destructive interference in terms of the phase difference between two waves are specified as follows:

- Two waves interfere constructively if at the point of superposition they are in phase, *i.e.*, the phase difference ϕ between them is $n \times 2\pi$ where *n = 0, 1, 2, ...*

- Two waves interfere destructively if at the point of superposition they are completely out of phase, *i.e.*, the phase difference ϕ between them is $(2n+1) \times \pi$.

These conditions are shown graphically in Figure 6.4. For any other phase difference, the intensity shall lie anywhere between the two extreme cases, *i.e.* for $\phi \neq n \times 2\pi$ and $\phi \neq (2n+1) \times \pi$, $I_{max} > I > 0$.

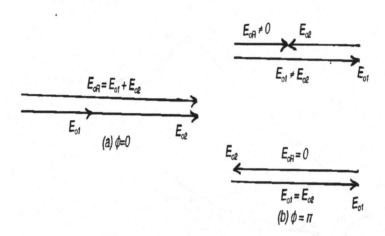

Figure 6.4: Graphical representation of (a) constructive, and (b) destructive interference of two waves of equal and unequal amplitudes.

Conditions of interference can also be expressed in terms of the optical path difference, Δx, between the two rays which is the difference between the optical paths traversed by the waves from their respective points of origin to the point of superposition. The optical path of a ray in a medium of refractive index n is given by $n \times physical \quad path \quad length$, and represents the distance traveled by the ray in vacuum within the same time interval. One complete wavelength (λ) corresponds to a phase angle of 2π. From this equivalence the conditions of interference in terms of path difference can be stated as follows.

- Two waves interfere constructively if at the point of superposition, the optical path difference Δx between them is an integer multiple of wavelength, *i.e.*, $\Delta x = 0, \lambda, \ldots, n\lambda$

- Two waves interfere destructively if at the point of superposition, the optical path difference Δx between them is an odd-integer multiple of half-wavelength, *i.e.*, $\Delta x = \lambda/2, 3\lambda/2 \ldots, (2n+1)\lambda/2$.

6.4.1   Phase Change on Reflection and Conditions of Interference

Without proving explicitly, when a light wave traveling in a rare medium is reflected from a surface backed by a denser medium, the phase of the reflected wave changes by π. This corresponds to an additional optical path diference of $\lambda/2$. As a result the conditions of interference between the incident and the reflected waves must be reconsidered. In terms of phase difference the conditions of interference remain the same, provided one takes into consideration the extra phase difference of π

introduced in the reflected wave. In terms of the optical path difference the conditions of interference are reversed as given below.

- An incident wave and a wave reflected from the surface of a denser medium interfere constructively if at the point of superposition optical path difference Δx between them is an odd-integer multiple of half-wavelength, *i.e.*, $\Delta x = \lambda/2, 3\lambda/2 \ldots, (2n+1)\lambda/2$.

- An incident wave and a wave reflected from the surface of a denser medium interfere destructively if at the point of superposition the optical path difference Δx between them is an integer multiple of wavelength, *i.e.*, $\Delta x = 0, \lambda, \ldots, n\lambda$

6.5 Young's Double Slit Experiment

The double slit experiment by Thomas Young in 1801 is the simplest way to demonstrate and study interference of light in the laboratory. Initially it was not accepted as a convincing evidence of the wave nature of light, and faced objections from most members of the scientific community, the objection being that the fringes could have been produced due to some *modification* of light by the edges of the slits, and not due to interference. Later, the doubts and objections having been resolved, the double slit experiment is historically the first demonstration of the wave nature of light. The principle of the experiment was extended to study the wave nature of particles such as electrons. In a 2002 survey amongst the readers of Physics World to which 200 Physicists had responded, the interference of single electrons by Young's double slit experiment was in fact voted to be the *most beautiful experiment in Physics*.

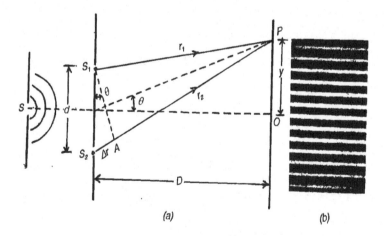

(a)   *(b)*

Figure 6.5: (a) Experimental setup for the Young's double slit interference. (b) Young's double slit interference pattern (schematic).

Figure 6.5 (a) shows the experimental arrangement of Young's double slit interference experiment. The two coherent beams of light to produce interference are obtained by the division of wavefront. Monochromatic light of wavelength λ from a narrow horizontal slit S produces cylindrical wavefronts which are incident on a pair of narrow, horizontal slits S_1 and S_2 distance d apart in a opaque screen, placed symmetrically parallel to the source slit S. Separation of the source slit and the double slits is immaterial, and it will only effect the brightness of the interference pattern. The double slits act as a pair of coherent sources, and light waves from them superimpose to give a stable interference pattern of horizontal dark and bright fringes on a screen placed at a distance D parallel to the double slits. The interference pattern can either be photographed by replacing the screen with a photographic plate, or it can be taken on a ground glass plate, and analyzed using a microscope from behind the screen. Figure 6.5 (b) shows a typical Young's double slit interference pattern.

At point O equidistant from the two slits, waves from both slits arrive in phase, and interfere to give a maxima of intensity. At a general point P at angular position θ the mode of interference shall depend on the phase difference between the two waves, which in turn depends on their path difference: $\Delta r = |r_1 - r_2| = d\,sin\theta$. The interference at point P shall be constructive, and the interference fringe shall be bright, if:

$$\Delta r = d\,sin\theta = m\lambda; \quad m = 0, \pm 1, \pm 2, \ldots, \tag{6.8}$$

and the fringe at point P shall be dark if:

$$\Delta r = d\,sin\theta = (2m+1)\frac{\lambda}{2}; \quad m = 0, \pm 1, \pm 2, \ldots, \tag{6.9}$$

where m is known as the order of the bright/ dark interference fringe. Equations (6.8 and 6.9) can be used to obtain the transverse location of fringes along the screen measured from the zeroth order bright fringe at O. Let $OP = y$. Then from Figure 6.5 (a), for small angle θ:

$$y = D\,tan\theta \approx D\,sin\theta. \tag{6.10}$$

Combining equation (6.10) with equations (6.8) and (6.9) respectively, we obtain:

$$y_{m,bright} = \frac{\lambda D}{d}m \tag{6.11}$$

and

$$y_{m,dark} = \frac{\lambda D}{d}\left(m + \frac{1}{2}\right) \tag{6.12}$$

From equations (6.11 and 6.12), separation $\Delta y_{bb}, (\Delta y_{dd})$ between consecutive bright (dark) fringes is given by:

$$\Delta y_{bb} = \Delta y_{dd} = \frac{\lambda D}{d} \tag{6.13}$$

and the separation $\Delta y_{bd}(\Delta y_{db})$ between adjacent bright and dark fringes is given as:

$$\Delta y_{bd} = \Delta y_{db} = \frac{1}{2}\frac{\lambda D}{d} \tag{6.14}$$

From equations (6.11 to 6.14) the following features of Young's double slit interference pattern are note worthy:

- Fringe width is proportional to the wavelength of light. Fringes for longer wavelength λ_l, say red light are wider than the fringes for shorter wavelength λ_s say blue. Thus, in the interference pattern of polychromatic light, m^{th} fringe of wavelength λ_l shall coincide with the $(m+n)^{th}$ fringe of wavelength λ_s.

- Fringe separation is proportional to the distance D between the double slits and the screen. A larger D gives wider fringes, but with the increase in D the intensity of light diminishes as square of the distance. Typically D is of the order of 1 m.

- Separation between the fringes is inversely proportional to separation d between the slits. The closer the slits are, the wider the fringes. Typically d is of the order of 1 mm.

- Since in a typical experimental set up, D/d is of the order of 10^3, the separation between the fringes is of the order of 10^{-4} m. One, therefore, requires a microscope to study and analyze the interference pattern.

6.5.1 Intensity Distribution in the Double Slit Interference Pattern

Intensity distribution in the interference pattern of two waves of equal amplitude (equations 6.1 and 6.2), was obtained as given by equation (6.7). For the case of Young's double slit experiment, the phase difference ϕ between the two waves interfering at point P (Figure 6.5) is:

$$\phi = \frac{2\pi}{\lambda}\Delta r = \frac{2\pi}{\lambda}d sin\theta \qquad (6.15)$$

Substituting $sin\theta = (y/D)$ from equation (6.10) in equation (6.15) gives:

$$\phi = \frac{2\pi}{\lambda}d\frac{y}{D} \qquad (6.16)$$

and substituting the expression for ϕ in the equation (6.7), the intensity distribution of the double slit interference pattern is given by:

$$I = I_{max}cos^2\left(\frac{\pi d}{\lambda D}y\right) \qquad (6.17)$$

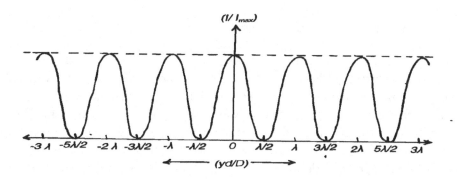

Figure 6.6: Simulated normalized intensity distribution of the double slit interference pattern.

Figure 6.6 shows schematically the normalized intensity distribution I/I_{max} as a function of the normalized distance $(d/D)y$ along the screen. The theoretical intensity distribution displays all the aspects of the pattern in regard to the separation of fringes discussed above. In addition, the intensity of all bright fringes are equal, and the intensity of dark fringes is zero.

In this section, we have dealt with the interference of light from the double slits completely ignoring the *diffraction* effect, which is discussed in Chapter 7. In reality, each of the slits produces a diffraction pattern of its own, and the two diffracted intensities superimpose to give a resultant interference pattern (see section 7.5).

6.6 Other Examples of Interference by Division of Wavefront

6.6.1 Fresnel Biprism Experiment

Around 1814 Fresnel produced interference of light using a biprism to divide the wavefront, and the interference pattern observed was not subject to the same objection as the double slit experiment. Thus biprism experiment at that time was the first experimental demonstration of the wave nature of light free from doubts and controversy, and bears a historical importance.

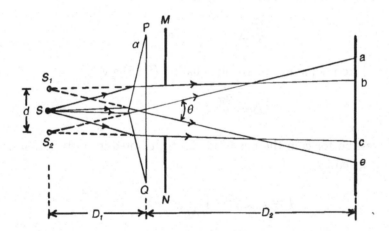

Figure 6.7: Experimental setup for the Fresnel's biprism interference experiment.

The biprism is a very thin glass prism of a very large prism angle ($\sim 179^o$) with the two prism surfaces almost along the same straight line. It can be easily produced in the laboratory by bavelling off by about 1° and polishing half portion of a microscope slide. Figure 6.7 shows the experimental set up to produce interference pattern by division of wavefront by the Fresnel biprism. The angle of prism shown is smaller than the actual for the clarity of drawing. A wavefront of monochromatic light from slit S on incidence on the prism is divided into two. Portion of the wavefront refracting through the upper part of the prism is deviated downwards, and the portion that passes through the lower portion of the prism deviates upwards. The refracted rays from two parts of the biprism appear to diverge from virtual images S_1 and S_2 in the same vertical plane as S which act as the two virtual coherent sources producing the interference pattern. An interference pattern of equi-spaced interference fringes is observed in region *bc* on the screen in which the refracted rays from two portion of the biprism superimpose. Screen *M* with a wide slit is used to block the rays from the edges of the biprism from reaching the observation screen, which produce diffraction pattern from the straight edges *P* and *Q* of the biprism on either side of the interference pattern.

Denoting the separation between the two virtual sources S_1 and S_2 by *d*, and $D = D_1 + D_2$ as the distance of the screen from slit *S*, the mathematical analysis of the interference pattern is the same as for the double slit experiment. It is left to students to show that the separation between the two consecutive bright (dark) fringes on the screen is the same as given by equation (6.13). The separation *d* between the virtual sources can be taken as: $d = D_1\theta$ to a very good approximation because the angle θ is very small, less than 0.05 radian. The angular separation θ can be measured by a spectrometer.

6.6.2 Fresnel's Mirrors

The Fresnel's mirrors consist of two plane mirrors inclined to each other at a very small angle ($\sim 1^o$). The incident wavefront from a slit *S* on reflection from the mirrors is divided producing virtual images S_1 and S_2 of the slit, which act as the virtual coherent pair of sources. Interference fringes are observed in area of the screen in which the reflected rays from both the mirrors overlap (Figure 6.8). The angular separation 2θ between S_1 and S_2 is twice the angle θ between the mirrors, and their distance from the mirrors is the same as that of slit *S*, from which the separation *d* between S_1 and S_2 can be calculated. Both portions of the wavefront on reflection under go a phase change of π, and thus remain in phase even after reflection. Thus, the conditions of bright and dark fringes are the same as for the double slit experiment, and so is the complete mathematical treatment of the problem.

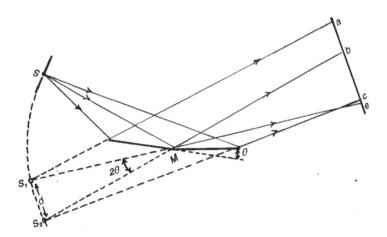

Figure 6.8: Interference of light by division of wavefront by Fresnel's mirrors.

6.6.3   Lloyd's Mirror

The Lloyd's mirror M comprises an ordinary glass plate 2 to 5 cm wide and 20 to 30 cm long, as illustrated in Figure 6.9.

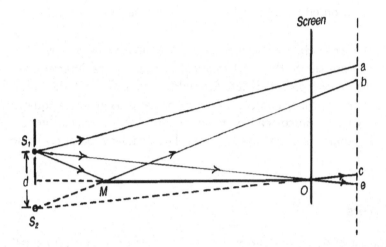

Figure 6.9: Lloyd's mirror experiment for the interference of light.

Light from slit S is reflected from the mirror at grazing incidence, and the slit S and its virtual image S' in the mirror act as the pair of coherent sources. The observation screen is placed in contact with the far end of the mirror. Interference pattern of equi-spaced dark and bright fringes are observed in the region of the screen in which the reflected rays from the mirror and the direct rays from the slit S overlap. As was stated in Section 6.4.1, reflected ray from the mirror undergoes a phase change of π, where as the direct rays from slit S remain unchanged. Therefore, in this case the conditions for bright and dark fringes are interchanged from those for the double slit experiment, and the conditions for destructive and constructive interference in terms of the optical path difference Δr of two rays in Lloyd's mirror experiment are:

$$\Delta r = dsin\theta = m\lambda; \quad m = 0, \pm1, \pm2, \ldots, \quad (Dark \;\; fringe) \quad\quad (6.18)$$

$$\Delta r = dsin\theta = (2m+1)\frac{\lambda}{2}; \quad m = 0, \pm1, \pm2, \ldots, \quad (Bright \;\; fringe) \quad\quad (6.19)$$

The fringe at the edge of the mirror, corresponding to the central, 0th order bright fringe in double slit interference pattern, is a dark interference fringe.

6.7   Interference by Division of Amplitude

Interference of light by division of wavefront of which a number of cases have been discussed in the previous section requires specialized equipment, and can only be observed under controlled laboratory conditions. On the other hand, interference of light by division of amplitude is a common phenomenon in nature, and almost everyone has the experience of having seen it. Examples of such observations are the beautiful multi-colours of a soap bubble, colours of thin film of oil on water or any other liquid, colours of cracks in a glass plate, colours of thin film coatings on optical components such as sunglasses or camera lenses, bright attractive colours of peacock feathers etc. Although the observer may not be aware of the cause of the observed colours, they are all produced from the interference of white light by thin films. The white light consists of the entire range of visible wavelengths that manifest as different colours in the spectrum. When thin films are observed in white light, the light incident at different angles and/ or at different thicknesses of the film satisfy the conditions of constructive interference for different colours which give rise to the observed colours. In order to consider interference by division of amplitude analytically and make measurements, we reinforce the following key facts:

- As stated in Section 6.4.1, a ray of light traveling in a rare medium on reflection from the surface of a denser medium undergoes a phase change of π radians. There is no change of phase on reflection from the surface of a rarer medium or on refraction.

- On entering a medium of refractive index n, the wavelength λ_o of light in vacuum changes to $\lambda_n = \lambda_o/n$. From this one defines the optical path of a ray in a medium of refractive index n as in Section 6.4.

- The amplitude of the incident wave on reflection changes by a factor r (< 1), coefficient of reflection (also known as reflectance) defined as:

$$r = \frac{amplitude \ \ of \ \ reflected \ \ wave}{amplitude \ \ of \ \ incident \ \ wave} \tag{6.20}$$

The reflectance at *normal incidence* for a medium of refractive index n, bounded by air with refractive index n_{air}, is given as:

$$r = \frac{n - n_{air}}{n + n_{air}} \tag{6.21}$$

- The coefficient of transmission (or transmittance of an interface) t is defined as:

$$t = \frac{amplitude \ \ of \ \ refracted \ \ wave}{amplitude \ \ of \ \ incident \ \ wave} \tag{6.22}$$

Thus, when a wave traveling in one medium enters another medium, its amplitude changes by a factor $t(< 1)$.

- The coefficients of reflection at normal incidence for an interface between two media for waves reflected from either side of the interface are equal in magnitude (equation 6.21). However, the coefficient of transmission for an interface between two media depends on the side from which the wave crosses the boundary. If the coefficient of transmission from medium 1 to medium 2 is t, then the coefficient of transmission from medium 2 to medium 1, $t' \neq t$. This can be proved from the reversibility of the path of light following Stokes treatment (Fig 6.10). Consider a wave of amplitude a partly reflected and partly transmitted at the interface of two media. The amplitude of the reflected wave is ar and that of the transmitted wave is at (Figure 6.10 a). Now reverse the directions of the reflected and transmitted waves. Both these waves are reflected and transmitted at the interface. If the coefficient of reflection in medium in 2 is r' and the coefficient of transmission from medium 2 to medium 1 is t' then the amplitudes of the reflected and transmitted waves resulting from the two incident waves at the interface are shown in Figure 6.10(b). Now from the superposition of reflected and transmitted waves in two media, and applying the reversibility of the path of light:

$$att' + arr = a, \tag{6.23}$$
$$art + atr' = 0 \tag{6.24}$$

Simplifying equations (6.23 and 6.24), we get:

$$tt' = 1 - r^2, \tag{6.25}$$
$$r' = -r \tag{6.26}$$

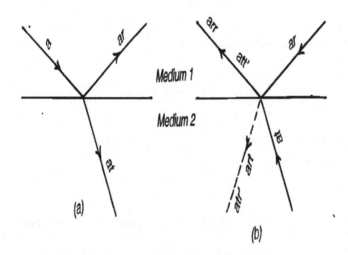

Figure 6.10: Stokes treatment of reflection and transmission at an interface.

- For a wave traveling within the same medium, the intensity is proportional to the amplitude of the wave. However for a wave traveling from one medium to another medium, the intensity of light besides being proportional to the square of the amplitude of the wave, also depends on the refractive index of the medium. Therefore, the conservation of energy of a wave traveling from one medium to another medium can not be expressed as $r^2 + t^2 = 1$. Hence, $t \neq t'$.

- When a wave confined in a medium is successively reflected and transmitted from the two boundaries of the medium, the amplitude of the wave on each reflection and transmission changes by a factor of r and t or t' respectively. This is shown in Figure 6.11. It is of course assumed that there is no absorption of energy by the medium.

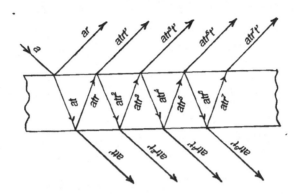

Figure 6.11: Amplitude of a wave on successive reflection and tranmission at the two boundaries of a medium.

6.7.1 Interference from a plane-parallel film

Consider a plane parallel film of thickness d of refractive index n with air (let $n_{air} = 1$) on both sides. A ray of light is incident on the top surface of the film at an angle ϕ, and undergoes multiple reflections and transmission on both surfaces of the film as shown in Figure 6.12. This gives a set of parallel rays on each side of the film. Since all these rays result from one single incident ray, they are coherent. When these parallel rays on either side are brought to focus by a lens, they superimpose to produce a maximum or minimum of interference depending on their relative phase differences. The conditions for bright and dark fringes can be worked out by calculating the optical path difference between successive rays, and by applying the conditions of constructive and destructive interference.

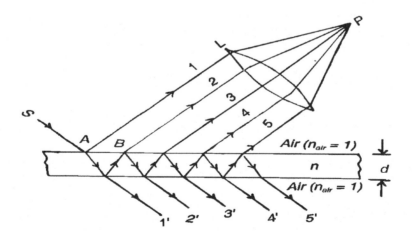

Figure 6.12: Interference from a plane-parallel film.

Interference on the incidence side

To calculate the optical path difference between successive rays *1* and *2*, consider Figure 6.13. Let ϕ' be the angle of refraction. *AC* and *BD* are perpendiculars to *BF* and ray *1* respectively. Optical paths of the two rays from *D* and *B* to the focus of the lens are equal, and the optical path difference between them is:

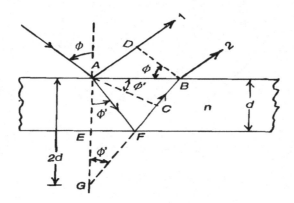

Figure 6.13: Optical path difference for derivation of conditions of interference from a plane-parallel film.

$$\Delta x = n \times (AFB) - AD = n \times (BFG) - AD = n \times (BC + CG) - AD \qquad (6.27)$$

But the optical path $AD = n \times BC$, (RE: Chapter 2, Section 2.4.1, refraction from Huygen's wave theory). Hence the net optical path difference between the two rays is:

$$\Delta x = n \times CG = n \times 2d\,cos\phi' = 2nd\,cos\phi' \qquad (6.28)$$

Since ray *1* on reflection undergoes phase change of $\pi (= path\ difference\ of\ \lambda/2)$, the two rays will interfere destructively resulting in a dark fringe if:

$$\Delta x = 2nd\,cos\phi' = m\lambda, \qquad (Dark fringe) \qquad (6.29)$$

and they will interfere constructively to produce a bright fringe if:

$$\Delta x = 2nd\,cos\phi' = \left(m + \frac{1}{2}\right)\lambda, \qquad (Bright fringe) \qquad (6.30)$$

where λ is the wavelength of the incident monochromatic light, and $m = 0, 1, 2, 3, ...$ If the incident ray strikes the upper surface of the film nearly normally, $\phi' \sim 0$, $cos\phi' = 1$, and the path difference $\Delta x = 2nd$, and similar conditions of destructive and constructive interference apply as given by equations (6.29) and (6.30) respectively.

Now let us consider other rays, numbered 3, 4, 5,... in Figure 6.12. From the figure we note the following important features of these rays.

- Since these rays are produced from reflections from the lower surface of the film backed by a rare medium, none of these rays undergoes a phase change of π on reflection as in the case of ray 2 which is reflected from the upper surface of the film.

- Optical path difference between each successive pair of rays, such as pair 2, 3, pair 3, 4, etc. is the same as between the pair of rays 1, 2, *i.e.* $\Delta x = 2nd\,cos\phi'$.

Let us first consider the destructive interference. If the condition of interference given by equation (6.29) is satisfied for rays *1* and *2*, then it will also be satisfied between the pairs of rays 1-3; 1- 4; 1-5; ..., because the path difference between these pairs of rays is an integer multiple of $m\lambda$. Thus each of the rays $2, 3, 4, 5, 6, ...$ will interfere destructively with ray 1. In other words, rays $2, 3, 4, 5, 6...$ are all in phase with each other, they together will interfere constructively, and the resultant wave then interferes with ray *1* destructively. The resultant amplitude *A* of the wave obtained from addition of the waves corresponding to rays 2, 3, 4, ... is:

$$\begin{aligned} A &= atrt' + atr^3t' + atr^5t' + atr^7t' + ... \\ &= atrt'(1 + r^2 + r^4 + r^6 + ...) \end{aligned} \tag{6.31}$$

Since *r* is less than *1*, the geometric series in equation (6.31) is convergent, and its sum is $1/(1-r^2)$, so that equation (6.31) becomes:

$$A = atrt'\frac{1}{1-r^2} \tag{6.32}$$

Using $tt' = (1-r^2)$ from equation (6.25), the amplitude of the resultant wave from the constructive interference of rays $2, 3, 4, 5, 6...$ from equation (6.32) is: $A = ar$, which is the same as the amplitude or ray 1. Hence the ray *1* interferes with rays $2, 3, 4, 5, 6...$ perfectly destructively, and the resulting minimum is a perfect dark fringe.

Now consider the constructive interference. If the condition given by equation (6.31) is satisfied by rays *1* and *2*, then by virtue of the same path difference and no phase change of π, pair of rays *2* and *3* shall be out of phase by π, and they shall interfere destructively. But the intensity of ray *2* is more than the intensity of ray *3*, and they do not cancel each other completely which results in a partly bright dark fringe. Similarly, pairs of rays $4 - 5; 6 - 7, 8 - 9; ...$ interfere destructively to produce partly bright dark fringes of successively decreasing intensity. Since stronger wave of each pair, namely rays $2, 4, 6, ...$ are in phase with ray *1*, the resultant intensity from each pair combines with ray *1* constructively to give a fringe brighter than the intensity of ray *1* itself.

Shape of fringes

The nature of fringes depends on the optical path difference, which is constant for a given value of the angle of refraction ϕ'. The locus of the point of constant ϕ' on the surface of the thin film is a circle about the center at the foot of perpendicular to the film from the source. Thus the thin parallel film fringes are concentric circular.

Interference of the transmitted rays

Now consider the interference of the emergent rays $1', 2', 3', 4', 5'...$ on the other side of the film. From the geometry of the system, optical path difference between the pair of rays $1'$ - $2'$ is the same as between the pair of rays 1 - 2, i.e., $\Delta x = 2nd\cos\phi'$, but for these rays there is no addition phase change of π. Hence, the conditions of their interference are:

$$2nd\cos\phi' = m\lambda, \qquad (Bright fringe) \tag{6.33}$$

$$2nd\cos\phi' = \left(m + \frac{1}{2}\right)\lambda, \qquad (Dark fringe) \tag{6.34}$$

From the above conditions it is to be noted that the zeroth fringe *(m = 0)* in the interference pattern of transmitted rays is a bright fringe, whereas on the incidence side of the film it is a dark fringe. Thus the interference patterns from the two sides of the film are complimentary to each other, *i.e.,* the sequence of dark and bright fringes in the two patterns are interchanged.

When the rays *1'* and *2'* are in phase, all the subsequent rays *3', 4', 5',...* shall also be in phase with them, and one gets a bright fringe from the constructive interference of these rays. When rays *1'* and *2'* are out of phase, the pair of rays *3'* - *4'*; *5'*- *6'*; ... are also out of phase. But the rays *1', 3', 5',...* being stronger than the rays *2', 4', 5',...*, they do not cancel each other completely. Thus the dark fringe is not a perfectly dark, but is partly bright, the brightness depending on the reflectance *r* of the film. If the reflectance is large, then the cancelation of intensity in destructive interference is higher, and the dark fringe is darker. On the other hand if the reflectance is small, then the amplitude of higher order rays drops rapidly, and only the first few rays are significant in the interference. The visibility of the interference does not depend on the absolute intensities of the dark and bright fringes, rather on their relative intensity. One defines the visibility *V* of the interference fringe pattern as:

$$V = \frac{I_{max} - I_{min}}{I_{max} + I_{min}}. \tag{6.35}$$

Thus the visibility of interference pattern is maximum, equal to one when the dark fringes are perfectly dark as is the case for the interference from the incidence side of the film. Lastly, when the zeroth order fringe from the incidence side of the film is perfectly dark, the entire intensity of the incident ray is transferred to the zeroth order bright fringe in the transmitted light, thus conserving the energy.

A general treatment of interference from a thin parallel film

In the preceeding section we considered only the special cases of constructive and destructive interference from a thin plane parallel film. A general case for any thickness and angle of incidence, not necessarily resulting in path difference of $m\lambda$ or $(m + 1/2)\lambda$ is dealt with in terms of complex amplitudes of the reflected (transmitted) waves.

Consider the reflected and transmitted waves from the upper and lower surfaces of a thin plane film of thickness *d* and refractive index *n*. Let *r* and *t* be the coefficients of reflectance and transmittance respectively from the upper surface, and *r'* and *t'* are the corresponding coefficients from the lower surface as defined in relation to Figure 6.10. From equations (6.25) and (6.26), $tt' = 1 - r^2$ and

$r' = -r$. The phase difference 2δ between two successive waves numbered *1* and *2* in Figure 6.12, corresponding to the path difference given by equation (6.28) can be expressed as:

$$2\delta = \frac{2\pi}{\lambda} \Delta x = \frac{2\pi}{\lambda} 2nd\cos\phi \qquad (6.36)$$

If one has a wave of amplitude A with a phase factor of α, then its amplitude and phase factor are expressed in terms of a complex amplitude as $Ae^{-i\alpha}$, and the amplitude of the wave is the real part of the complex amplitude, which in this case is $A\cos\alpha$. Using the similar notations and referring to Figures (6.11) and (6.12), the complex amplitudes of the reflected-transmitted waves *1, 2, 3, ...* from the upper surface of the film are given as:

$$ar, \; att'r'e^{-2i\delta}, \; att'r'^3e^{-4i\delta}, \; att'r'^5e^{-6i\delta}, \; att'r'^7e^{-8i\delta}, \; att'r'^9e^{-10i\delta}... \qquad (6.37)$$

where *a* is the amplitude of the incident wave. Likewise, the complex amplitudes of the successive transmitted waves $1', 2', 3', ...$ are given as:

$$att', \; att'r'^2e^{-2i\delta}, \; att'r'^4e^{-4i\delta}, \; att'r'^6e^{-6i\delta}, \; att'r'^8e^{-8i\delta}, \; att'r'^10e^{-10i\delta}... \qquad (6.38)$$

Now considering the reflected-transmitted waves from the upper surface. The total amplitude (complex) is obtained by the summing up all the reflected-transmitted amplitudes (equation 6.37).

$$
\begin{aligned}
A_R &= ar + att'r'e^{-2i\delta} + att'r'^3e^{-4i\delta} + att'r'^5e^{-6i\delta} + att'r'^7e^{-8i\delta} + att'r'^9e^{-10i\delta}... \\
&= ar + att'r'e^{-2i\delta}(1 + r'^2e^{-2i\delta} + r'^4e^{-4i\delta} + r'^6e^{-6i\delta} + r'^8e^{-8i\delta}...) \\
&= ar + att'r'e^{-2i\delta}\left(\frac{1}{1 - r'^2e^{-2i\delta}}\right)
\end{aligned}
\qquad (6.39)
$$

Using $r' = -r$ and $tt' = (1 - r^2)$ in equation (6.39), and after simplifying one gets:

$$A_R = ar\left(\frac{1 - e^{-2i\delta}}{1 - r^2e^{-2i\delta}}\right) \qquad (6.40)$$

The total reflected amplitude, say A_{RR} is given by the real part of A_R in equation (6.40), and is obtained as follows:

$$
\begin{aligned}
A_{RR} &= Re(A_R) = Re\left\{ar\left(\frac{1 - e^{-2i\delta}}{1 - r^2e^{-2i\delta}}\right)\right\} \\
&= ar\left\{\frac{Re\{(1 - e^{-2i\delta})(1 - r^2e^{2i\delta})\}}{(1 - r^2e^{-2i\delta})(1 - r^2e^{2i\delta})}\right\} \\
&= ar\frac{(1 + r^2)((1 - \cos 2\delta))}{(1 + r^4 - 2r^2\cos 2\delta)}
\end{aligned}
\qquad (6.41)
$$

The intensity of the reflected light is simply the square of the reflected amplitude, *i.e.*, $I_R = A_{RR}^2 = I_0\mathcal{R}$ where I_0 is the incident intensity and $\mathcal{R}$ is the reflection coefficient of the film. The special cases of constructive and destructive interference are obtained from equation (6.41) be setting the phase difference 2δ to appropriate values. If $2\delta = 0, 2\pi, 4\pi, ...$ the reflected amplitude from equation (6.41), $A_{RR} = 0$ and one gets a perfectly dark fringe, and if $2\delta = \pi, 3\pi, ...$, one gets a bright fringe of amplitude $A_{RR} = (2ar)/(1 + r^2)$. These are the same results as we obtained in the preceeding section from a simple direct treatment of special cases only. How are the phase difference considered here related to the conditions of constructive and destructive interference stated in terms of path difference for the special cases? A similar treatment of the transmitted waves using complex amplitude is left as an exercise for students.

Thin film interference with white light

For the sake of discussion, we shall consider the interference pattern seen from the incidence side of the film, and a similar discussion shall apply to the interference seen from the other side of the film. Because the refractive index depends on wavelength , waves of different wavelength (colours) in white light are refracted into the film at different angles ϕ'. The optical path difference for the two extreme colours, (violet: V; red: R) of spectrum are:

$$\Delta x_V = 2n_V d\,cos\phi'_V = 2d\frac{sin\phi}{sin\phi'_V}\,cos\phi'_V \qquad (6.42)$$

$$\text{or } \Delta x_V = 2d\,sin\phi\,cot\phi'_V \qquad (6.43)$$

$$\text{and } \Delta x_R = 2d\,sin\phi\,cot\phi'_R \qquad (6.44)$$

Since $n_V > n_R$ for any medium, the angle of refraction $\phi'_V < \phi'_R$, and $\Delta x_V > \Delta x_R$. Consequently, different colours of spectrum satisfy the conditions of constructive and destructive interference at different points, and each colour of spectrum produces its own interference pattern of concentric dark and bright fringes which overlap. For the zeroth order minima ($m = 0$), all colours satisfy the condition of destructive interference simultaneously, and the first minima of each colour coincide. Thus the first fringe is a dark fringe. Beyond that the fringes of each colour are displaced with respect to each other, and the first bright fringe is not a pure white bright fringe, but it is composed of coloured fringes staring with the violet colour fringe first and the red colour fringe last. Moving further along the interference pattern the overlap of various colour fringe patterns becomes more complex; the bright fringes of one colour overlaps with the different order dark fringes of other colours, or the mth order bright fringe of one colour overlaps with the nth bright fringe of another colour. Thus, after a first few colour fringes, the interference pattern turns into a haze of colours, and eventually no distinct interference pattern is seen. Thin film interference with white light also depends on the thickness of the film. If the thickness of the film is large, the condition of maxima for every point of the film will be satisfied for a number of wavelength for different values of the integer m. Thus every point of the film, including the points of minima in monochromatic light, appear as bright fringe of a mixture of colours. As the thickness further increases, the number of colours producing bright fringes from each point increase, and the film will appear uniformly bright white. This explains why colour patterns are not seen in the thick layer of oil on water in the same way as one sees them in a thin film.

This phenomenon explains the observed colours of soap bubbles and of thin film of oil on water surface. Since the thickness of the bubble wall, and that of the oil film is not uniform throughout, the colour interference pattern is too complex to be treated analytically. All one sees is the profusion of luxuriant colours from different parts of the film which may be the bright interference fringes of those colours or absence of complimentary colours from white light due to corresponding dark fringes.

Application: Anti-reflection coating

This is one of the most important and widely used application of thin plane parellel films based on interference. High quality optical components such as camera lenses, microscope and telescope lenses, and solar cells in renewable energy applications are coated with a very thin anti-reflection film. The purpose of the film is to minimize the loss of incident light by minimizing reflection from the incident surface so that most of the light is transmitted through the lens or is received by the solar cell.

The coating is of a material of refractive index n' which is less than the refractive index of the material of the lens (or solar cell). As a result rays reflected from both the surfaces undergo a phase change of π, which cancel out, and the condition for the destructive interference for these rays at normal incidence is:

$$2dn' = \left(m + \frac{1}{2}\right)\lambda \tag{6.45}$$

where $m = 0, 1, 2, 3, ...$ and λ is the wavelength of light. For the zeroth order minima $m = 0$, and thickness of the film must be:

$$d = \frac{\lambda}{4n'} \tag{6.46}$$

For the destructive interference to be complete, the fraction of amplitude reflected at each of the two surface must be exactly the same. The reflection coefficients of the two surfaces in terms of the refractive indices of the media separated by the surfaces are given by:

$$r_1 = \frac{n' - 1}{n' + 1} \tag{6.47}$$

$$r_2 = \frac{n - n'}{n + n'} \tag{6.48}$$

where we have assumed $n_{air} = 1$. Equating equations (6.47) and (6.48) for equal refractive index of the two surfaces give:

$$n' = \sqrt{n} \tag{6.49}$$

Thus the refractive index of the anti-reflection coating must satisfy equation (6.49), and its minimum thickness must be equal to the quarter wavelength of light in the coating material. Other allowed thicknesses of the film are $(3\lambda/4n'), (5\lambda/4n'), (7\lambda/4n'), ..$ corresponding to $m = 1, 2, 3, ...$ For white light one uses an average value of wavelength. Lastly, there is no loss of energy due to antireflection coating. Energy removed from the reflected beam by destructive interference is redistributed in the transmitted beam.

6.7.2 Interference from a wedge shaped film

A wedge shaped film is produced from two plane glass plates, such as microscope slides, placed in contact with each other at one end and separated by a thin paper or a thin uniform wire at the other end (Figure 6.14). A medium of refractive index n (could also be air) is introduced between the glass plates. Let θ (radians) be the angle of the wedge-film. A monochromatic ray of light of wavelength λ incident on the upper glass plate is reflected from the surfaces of the glass plates that constitute the upper and lower surfaces of the wedge as shown by the odd numbered rays 1, 3, 5,... and even numbered rays 2, 4, 6,... respectively.

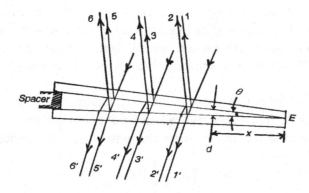

Figure 6.14: Interference from a wedge shaped film.

If d is the thickness of the film, at a lateral distance x from the edge of the wedge, then $d = x\tan\theta \sim x\,\theta$, and the optical path difference between the pair of rays reflected from the two surfaces of the film (for example rays *1* and *2*) is $2\,n\,d$, provided the observation is made almost normally to the film. The ray which is reflected from the lower surface of the film undergoes a phase change of π. Therefore the two rays interfere constructively to produce a bright fringe if:

$$2nd = 2nx\theta = \left(m + \frac{1}{2}\right)\lambda \qquad (6.50)$$

and they will interfere destructively resulting in a dark fringe if:

$$2nd = 2nx\theta = m\lambda \qquad (6.51)$$

where $m = 0, 1, 2, 3, ...$ Rays from the two surfaces of the film are not parallel, but appear to diverge from a point within the film, and the fringes appear to be formed in the film itself. Such fringes are known as *localized fringes*.

At the edge of the film, the thickness of the film is zero, and the condition of destructive interference is satisfied. At this point one sees a dark fringe of zeroth order followed by a pattern of bright and dark fringes. The dark fringes are not perfectly dark because the intensity of the interfering waves are unequal. The locus of the same thickness of the film is a straight line parallel to the edge of the film. Therefore, the fringes are straight lines parallel to the edge of the film. The lateral separation Δx between two successive bright or dark fringes is given by:

$$2n\Delta d = \lambda = 2n\Delta x\,\theta \qquad (6.52)$$

or

$$\Delta x = \frac{\lambda}{2n\,\theta} \qquad (6.53)$$

Thus the fringes are equally spaced.

Lastly, if white light is used, all colours satisfy the condition of a dark fringe at the edge of the film, and the edge is again a dark fringe. After that the condition of bright fringe for the shortest wavelength, *i.e.*, violet is satisfied, and the first bright fringe is violet followed by other colour fringes. Red with the longest wavelength is the last bright fringe in the series. After only a few distinct colour fringes, the interference pattern due to different wavelengths begins to overlap, and within a short distance from the edge the pattern changes first to a hue of colours, and then to nearly uniform white light at large thickness of the wedge. Why?

Interference in transmitted light

Interference can also be seen in the rays transmitted from the other side of the film, such as rays *1'* and *2'* in Figure 6.14. Ray *1'* is the straight transmitted ray without any reflection whereas ray *2'* undergoes two reflection from the glass surfaces and its phase changes by 2π. Therefore, rays *1'* and *2'* are in phase, and the conditions of their constructive and destructive interference are reverse of those given by equations (6.50) and (6.51) respectively. Thus, in the transmitted waves the edge of the film is a bright fringe followed by alternate dark and bright fringes, and the interference pattern is complimentary to the pattern seen from the upper surface of the film. The dark fringes, once again, from the upper surface of the thin film, are not perfectly dark, rather partly bright.

6.7.3 Newton's Rings

Newton's rings is the simplest practical application of interference by division of amplitude that can be applied for a number of optical measurements such as unknown wavelength, refractive index of fluids, difference in wavelength of two close spectral lines. The equipment needed are an optically flat glass plate, a convex lens, a collimated source of light, a microscope slide, a traveling microscope, and a piece of black paper. The experimental setup is shown in Figure 6.15.

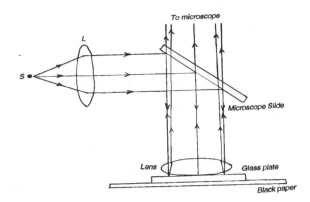

Figure 6.15: Experimental setup for Newton's Rings.

When a convex lens is placed on a plane glass plate, a circular-wedge shaped film of air (or of fluid introduced between them) is formed. The surface of the film in contact with the lens is spherical, and follows the curvature of the lens surface. Waves reflected from the upper and lower surfaces of the circular-wedge interfere in the same way as in the case of a plane wedge shaped film (Section 6.7.2, and follow the conditions of bright and dark fringes given by equation (6.50) and (6.51). Since the locus of the same thickness of the film is a circle with center at the point of contact between the lens and the plate, the interference pattern comprises concentric circular dark and bright localized fringes. The central spot, the point of contact between the lens and the glass plate is a dark fringe. Why? If the lens and the glass plate are not clean, and there is a speck of dust between them at the point of contact, the central spot may be a bright fringe. Explain why.

To relate the thickness d of the film to the diameter D_m of the mth dark (bright) interference fringe, consider Figure 6.16. Let R be the radius of curvature of the surface of the lense in contact with the glass plate. From the figure:

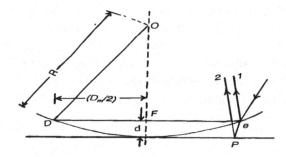

Figure 6.16: Path difference of two rays producing Newton's Rings.

$$FD^2 = OD^2 - OF^2 \Rightarrow \left(\frac{D_m}{2}\right)^2 = R^2 - (R-d)^2 = 2Rd - d^2 \qquad (6.54)$$

If $d \ll R, d^2 \lll 2Rd$, and equation (6.54) gives: $(D_m)^2 = 8Rd$.

From equations (6.50) and (6.51) the conditions of bright and dark Newton's rings in terms of their diameters are:

$$(D_m)^2 = 8Rd = \frac{4R}{n}(2nd) = \frac{4R}{n}\left(m + \frac{1}{2}\right)\lambda, \qquad (m^{th} \quad bright \quad rings) \qquad (6.55)$$

$$(D_m)^2 = 8Rd = \frac{4R}{n}(2nd) = \frac{4R}{n}m\lambda, \qquad (m^{th} \quad dark \quad rings) \qquad (6.56)$$

where n is the refractive index of the film, and $m = 0, 1, 2, 3, ...$ Generally it is not convenient, and advisable to start counting the rings from the center of the interference pattern, because one does not precisely know the order of the central fringe. Why? Instead one measures the diameters of clearly visible bright rings (Why not dark ?) let us say numbered l and $l+k$. Then from equation (6.55) (or 6.56 if one unwittingly measures the dark fringes):

$$D_{l+k}^2 - D_l^2 = \frac{4R}{n}k\lambda. \qquad (6.57)$$

Newton's rings of two close spectral lines

As an example consider the two close yellow spectral lines of sodium of wavelengths λ_1 (=589.592 nm) and λ_2 (=588.995 nm). Both the wavelengths produce their own overlapping pattern of Newton's rings. From equation (6.53) rings for the larger wavelength are slightly wider than those for the smaller wavelength, and the two rings patterns are slightly displaced with respect to each other. At the center, where the thickness of the film is zero, the zeroth order minima for both wavelengths coincide. As one moves outwards from the center, the two ring patterns begin to shift, and at some point the mth bright fringe of larger wavelength λ_1 shall coincide with the $(m+1)^{th}$ dark fringe of the shorter wavelength λ_2. At this point the fringes shall be least distinct. From the conditions of interference:

$$D_m^2 = \frac{4R}{n}\left(m + \frac{1}{2}\right)\lambda_1 \qquad (6.58)$$

$$D_{(m+1)}^2 = \frac{4R}{n}(m+1)\lambda_2 \qquad (6.59)$$

But from the overlap of fringe patterns, $D_m^2 = D_{(m+1)}^2$. Therefore, from equations (6.58) and (6.59):

$$\Delta\lambda = \frac{\lambda_2}{(2m+1)} \approx \frac{\lambda_{average}}{(2m+1)} \qquad (6.60)$$

For the sodium lines, $\Delta\lambda = 0.597 nm$ and $\lambda_{average} = 589.294 nm$. This gives the order of the fringes at the first location of least clarity as ≈ 490. In Newton's rings experiment one does not see as many fringes. Depending on the curvature of the lens, one is able to count only less than 100 fringes. But in principle, using Newton's Rings one can determine wavelength difference of two spectral lines, if the difference is of the order of $1 - 2\%$ of the average wavelength. Smaller wavelength difference, as for the sodium lines is determined using Michelson interferometer discussed in a later section.

Newton's rings in transmitted light and with white light

The discussion here remains the same as for the wedge shaped film. In transmitted light the central spot is bright, and the pattern is complimentary to the pattern seen from the other side. With white light, the central spot is dark, followed by colour rings starting with violet first, and then there will be an overlap of various colour bright and dark fringes, resulting in coloured hue and finally at sufficient distance it will be just white light.

Applications of Newton's Rings

1. Measurement of wavelength

The radius of curvature R of the lens surface is measured using a spherometer. For an air film $n = n_{air}$. From the measurement of the diameters of the rings, one can determine the unknown wavelength λ (from equation 6.57).

2. Measurement of refractive index

In order to determine the refractive index of liquids, one introduces a drop of liquid between the lens and the glass plate, and uses a monochromatic source of known wavelength, such as a sodium lamp. The refractive index of the material of the wedge-film between the lens and the glass plate is determined from the measurements of diameters and using equation (6.57). The determination of refractive index of liquid can be further simplified so that one need not determine the radius of curvature of the lens, and one does not require the precise value of the wavelength.

First, measure the diameters of the $(l+m)^{th}$ and l^{th} rings in the presence of air. from equation (6.57):

$$D^2_{air(l+m)} - D^2_{air(l)} = \frac{4R}{n_{air}} m\lambda. \tag{6.61}$$

Next introduce a drop of liquid between the lens and the glass plate, and again measure the diameters of the same order Newton's rings so that:

$$D^2_{l+m} - D^2_l = \frac{4R}{n} m\lambda. \tag{6.62}$$

From the ratio of equations (6.61) and (6.62) one obtains:

$$n = \frac{D^2_{air(l+m)} - D^2_{air(l)}}{D^2_{l+m} - D^2_l} n_{air} \tag{6.63}$$

The advantage of this method of measuring the refractive index of a liquid is that it requires very small quantity, just a drop, of liquid, and can be used for very rare or expensive liquids. As a precaution, while measuring the diameters, the traveling microscope must be moved in one direction only to avoid backlash error of the traveling mechanism. It is to be noted that in Newton's rings experiment one measures the diameters of the rings and not the radii. Why?

There is yet another interesting aspect of introducing a liquid between the lens and the glass plate. If the refractive indices of the lens, liquid and the glass plate are such that $n_{lens} > n_{liquid} > n_{plate}$ then none of the rays undergoes a phase change of π, or if $n_{lens} < n_{liquid} < n_{plate}$ then both the rays undergo a phase change of π. In both cases, the conditions of constructive and destructive interference are reversed, and the central spot is a bright fringe.

6.7.4 Michelson Interferometer

The Michelson interferometer is one of the most precise, yet simple in principle interferometer in which two coherent sources are obtained by the division of amplitude. It can detect a fraction of a fringe shift, or a wavelength difference of the fraction of nm. It is most widely known amongst the scientific community for the famous Michelson-Morley experiment to detect the presence of an absolute medium *ether* rather than for the routine interference studies. Figure 6.17 shows the interferometer schematically.

A collimated beam of monochromatic light is incident at an angle at an optically parallel glass plate P_1 half silvered at the back, known as the *beam splitter*. The incident beam is split into two beams

1 and *2* of equal amplitude by partial reflection and transmission from the silvered face. There in no change of phase of beam *1* due to reflection. Why? M_1 and M_2 are two front silvered mirrors of high optical quality mounted with their reflecting surfaces perpendicular to each other, facing plate P_1. Mirror M_1 is mounted on a precision screw carriage, and can be moved forward or backwards keeping perfectly parallel to itself. The movement can be measured up to a micrometer. Mirror M_2 can be tilted with screws mounted at the back so that it can be set perpendicular to M_1, or at an angle as may be required for a particular experiment (application).

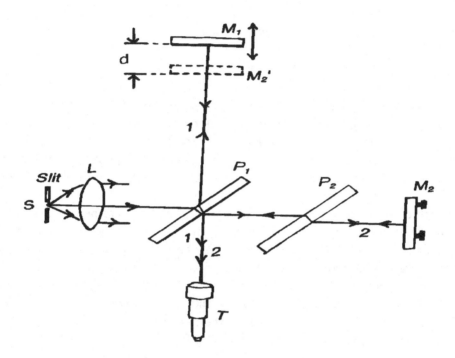

Figure 6.17: Michelson interferometer.

The two coherent waves *1* and *2* traveling perpendicular to each other from the beam splitter are incident on the mirrors normally, and reflected back to plate P_1. The waves are then directed by P_1 in to a telescope *T*, and superimposed to produce the interference pattern. Beam *1* directed towards mirror M_1 passes through the glass plate P_1 twice before it interferes with beam *1*, whereas beam *2* does not pass through the plate even once. To compensate for the optical path difference between the two beams introduced by plate P_1, a second optically identical glass plate P_2, known as *compensator*, is introduced parallel to P_1 in the path of beam *2*. The compensator plate is not an essential component of the interferometer as far as viewing of fringes is concerned, but it is indispensable for analytical work to nullify the effect of P_1 on the optical path difference of the two beams. With P_2 in place, the optical path difference between the two beams is simply equal to the physical path difference between the two beams measured from the point on P_1 at which the incident beam is divided.

Interference from Michelson Interferometer

In order to see an interference pattern from Michelson interferometer the following conditions must be met:

- The light source must be extended, and not a point source or a slit so that the entire area of the mirrors are filled with light. This is achieved by one of many possible ways which include using a large sodium or mercury lamp, or by introducing a lens between a point source and the interferometer such that the source is at the focal point of the lens, or by simply introducing a ground glass plate in front of the source.

- The light must be monochromatic or nearly so, for example the sodium light. This is particularly important if the distances of the mirrors from the beam splitter are not nearly equal.

1. Circular fringes

In the most common mode of the use of Michelson interferometer, the two mirrors are made perpendicular to each other with the help of screws at the back of mirror M_2, and their distances from P_1 are made nearly equal, to within millimeters by moving mirror M_1 in the required direction. Once these conditions have been fulfilled, a circular interference fringe pattern pops up in the telescope. In order to analyze the interference pattern, and to understand formation of the fringes, consider Figure 6.18.

M_2' is the virtual image of mirror M_2 reflected in P_1. Since M_1 and M_2 are perpendicular to each other, and if d is the difference between their distances from P_1, then M_1 and M_2' are parallel to each other separated by a distance d. The space between M_1 and M_2' acts like a thin plane parallel film of air of thickness d, and the interference between the rays reflected from the two mirrors is the same as if the two rays were reflected from the two surfaces of a plane parallel thin film of thickness d and refractive index $n = n_{air} = 1$. If θ is the angle at which the rays are reflected from the mirrors, then the path difference of the two rays is $2d\,cos\theta$ as in the case of a plane parallel film. Both rays undergo a phase change of π on reflection from the mirror surfaces.

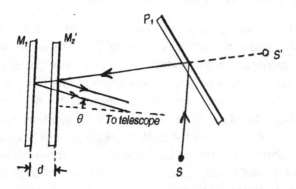

Figure 6.18: Optics of circular fringes in Michelson interferometer.

Thus the change in phase of both the rays due to reflection cancel out, and the conditions of interference are:

$$2d\,cos\theta \;=\; m\lambda, \qquad (Bright \quad fringes) \tag{6.64}$$

$$2d\,cos\theta \;=\; \left(m + \frac{1}{2}\right)\lambda, \qquad (Dark fringes) \tag{6.65}$$

where $m = 0, 1, 2, 3, ...$ Locus of the same θ is a circle, so that the alternate dark and bright fringes are concentric circles. The reflected rays are parallel that appear to converge to infinity and the fringes are formed at infinity. A special case arises when the distances of both the mirrors are optically equal giving *d = 0*, which is undefinable for a physical thin parallel film of refractive material. In this case the condition of constructive interference is satisfied, and the field of view appears to be filled with uniform intensity. If white light is used, we shall also get uniform white intensity in the field of view if *d = 0* or if $d >> \lambda_{visible}$. Why? But if *d* is small, comparable to the average visible wavelength, a few colour circular fringes close to the center shall be observed, followed by haze of colours and then uniform white light resulting from the over lap of many bright fringes of different colours at each point.

2. Localized fringes

If the two mirrors are not perfectly parallel, and are at small angle to each other, the 'virtual' thin film between M_1 and M_2' is wedge shaped. The interference pattern seen is the same as for a wedge shaped film of refractive index equal to 1, and the fringes are localized. If the angle between the mirrors is small, and for small thickness of the film fringes appear straight parallel to the edge of the wedge. If the angle between the mirrors is large, or at large thickness of the air film, the fringes are curved, convex towards the edge of the wedge, because the variation in the path difference with angle θ.

Applications of Michelson Interferometer

Applications of Michelson interferometer are many and varied. Below, we shall discuss just a few common ones plus historically important applications.

1. Measurement of wavelength

Consider a thickness *d* of the 'virtual' plane parallel air film and concentrate on the mth fringe at angular position θ. Now by moving mirror M_1 if *d* is decreased, $cos\theta$ must increase so that the product $2d\,cos\theta = m\lambda$ for the mth fringe remains constant. This implies that angle θ and hence the diameter of the fringe decreases. Thus, the mth fringe appear to shrink in diameter and move towards the center of the circle, and eventually it sinks into the center. Following the same logic, if *d* is increased, the fringes appear to expand and emerge from the center of the interference pattern. Since, the optical path difference between two consecutive bright or dark fringes is λ, for a distance $d/2$ moved by the mirror, one fringe will appear to sink into or emerge from the center of the interference pattern. In order to measure the wavelength, one slowly moves mirror M_1 either forward or backward by a distance δ and counts the number *N* of fringes disappearing into or emerging from the center. Then the wavelength of light is: $\lambda = 2\delta/N$.

2. Measurement of a small wavelength difference

As discussed in subsection 6.7.3, when interference patterns of two close wavelengths are formed simultaneously, they are displaced with respect to each other. As a result, at certain point in the interference pattern of both the wavelengths, a bright fringe due to one wavelength shall coincide with the bright fringe of another wavelength, and the fringes are said to be in consonance. At this point the fringes are most distinct. Likewise at another point a bright fringe of one wavelength coincides with the dark fringe of another wavelength, and the fringes are said to be in dissonance. The fringes at this point are least distinct. The points of consonance and dissonance occur alternately in the interference pattern. In going from one point of consonance to the next point of consonance, or from one point of dissonance to the next point of dissonance the fringe pattern of one wavelength shall shift with respect to that of the other wavelength by exactly one fringe, which in terms of optical path difference must be equal the mean wavelength. Let at some point of consonance at the center of the interference pattern ($\theta = 0$)(or dissonance) the 'air film' thickness is d_1. Then $2d_1 = m_1\lambda_1 = m_2\lambda_2$, or

$$2d_1 \left(\frac{1}{\lambda_1} - \frac{1}{\lambda_2} \right) = (m_1 - m_2) = an \quad integer \quad = k \qquad (6.66)$$

Now we move mirror M_1 till the next point of consonance appears at the center of the fringe pattern, and if d_2 is the thickness of the air film, then from equation (6.66)

$$2d_2 \left(\frac{1}{\lambda_1} - \frac{1}{\lambda_2} \right) = k + 1 \qquad (6.67)$$

From equations (6.66) and (6.67),

$$2(d_2 - d_1) \left(\frac{1}{\lambda_1} - \frac{1}{\lambda_2} \right) = 1 = 2\delta \frac{\Delta\lambda}{\lambda_1\lambda_2} \qquad (6.68)$$

This gives:

$$\Delta\lambda = \frac{\lambda^2}{2\delta} \qquad (6.69)$$

where $\lambda = \sqrt{\lambda_1\lambda_2}$ is the average wavelength, and δ is the distance moved by the mirror from once consonance to the next consonance or from one dissonance to the next dissonance.

3. Measurement of refractive index

When a thin plate of thickness t of refractive index n is introduced in the path of one of the beams, say beam 1 traveling to and from mirror M_1, the beam passes through the plate twice, and optical path of the beam changes by $2(n - n_{air})t$. This causes a shift of fringes in the interference pattern by Δm where m need not be integer. The shift in fringes is related to the change in optical path difference as:

$$2(n - n_{air})t = \Delta m \, \lambda \qquad (6.70)$$

from which the refractive index of the material can in principle be determined. However, practically it is not possible to determine the shift in fringes on introduction of the plate in the path of the beam, because fringes before and after are indistinguishable. This problem is overcome by introducing two identical plates of the material in both the beams, initially keeping both plates normal to the beam. Then one of the plate is gradually but slowly rotated about an axis in the plane of the plate and perpendicular to the beam and the shift in the fringes (emerging from or sinking into the center) is counted. Rotation of the plate changes the effective thickness of the plate in the path of the beam. Difference in the optical path of the two beams expressed in terms of *n, t* and the angle of rotation θ can be related to the shift in the fringes. As an exercise students are assigned to derive this relationship.

For the measurement of the refractive index of gases, gas is introduced slowly in to an evacuated tube placed in the path of one of the beam, and refractive index of the gas as a function of pressure is determined from the shift in fringes.

4. Standardization of the meter

The standard meter is the distance between two marks on a platinum-iridiun road at $0\ ^{o}C$ kept at the Bureau of Weights and Measures in Paris, and its replicas are kept at various other standards bureaus around the world. In the event of natural or man-made calamities, these standards are subject to destruction. The first non-destructible standard of meter length was established by Michelson and Benoit in late 1800s in terms of the wavelength of spectral lines of cadmium by interferometry.

In practice it is not possible to count the number of fringes displaced when the movable mirror is moved a whole meter length. The difficulty was overcome by using nine standard etalons (a specially designed assembly of two plane parallel mirrors separated by a distance *d*), and a Michelson interferometer of modified design. Each etalon was approximately twice the length of the previous one, the ninth being about *10 cm* long. First the number of fringes moved corresponding to the length of the shortest etalon was determined, and then by successively comparing each shorter etalon to the next longer one, the length of the ninth longest etalon in terms of wavelength was determined. Finally, the ninth etalon was compared to the meter length in ten steps to determine the length of the meter in terms of wavelength. The cumulative error of comparison of the last etalon to the meter rod in 10 steps was found to be less than the uncertainty in the end marks of the meter rod. In this way the meter length in terms of the three cadmium spectral lines was found to an accuracy of 1 in 10^{7}. These standards are given in Table 6.1 below:

Table 6.1: Standard meter length in terms of the wavelength of light.

Spectral Line	Wavelength (Å)	1 m =
Red Line of Cadmium, λ_R	6438.4722	1,553,163.5 λ_R
Green Line of Cadmium, λ_G	5085.8240	1,966,249.7 λ_G
Blue Line of Cadmium, λ_B	4799.9107	2,083,372.1 λ_B
Orange Line of Krypton, λ_O	6057.80211	1,650,763.73 λ_O

Interferometric measurements not only provided the non-destructible standard of the meter length, it also provided the accurate measurement of wavelengths in terms of the standard meter length, and the red cadmium line is the primary standard in spectroscopy. Later, measurement were made with the orange line of krypton in dry atmosphere at 15 ^oC and a pressure of 760 mm of Hg. These are also given in Table 6.1. At present the Krypton orange line is accepted as the international standard of length, adopted in Paris in October 1960.

5. Detecting the presence of ether

The mechanical waves such as the sound waves, and the waves in water require a medium to sustain them and to aid their propagation from one point to another in space. Following the discovery of electromagnetic wave nature of light, a medium called *ether* was postulated for the sustanance and propagation of electromagnetic waves. Ether pervaded the entire space including vacuum, was fixed with respect to fixed stars through which earth rotated, and endowed with infinite elasticity and inertia but zero mass. Michelson and Morley conducted a number of inteferometric experiments using Michelson interferometer to detect the presence of ether; the negative results of which negated the presence of ether and, lead to the birth of Einstien's theory of Relativity from which the much of modern physics as we know it today followed. It is for this historical importance we give some basic details of this famous experiment without going into the technical details that are included in the Modern Physics text books.

The motion of the earth through fixed ether shall cause ether wind for an earth observer. As a result two beams of light, one traveling parallel to the direction of the rotation of earth and another perpendicular to it shall have slightly different velocities. Michelson-Morley proposed to detect this difference of velocities by using Michelson interferometer that employs two perpendicular coherent beams for interference. The interferometer was placed on a massive solid rock of about $1m^3$ size floating in a pool of mercury to eliminate the effects of any small vibrations and disturbances. Interference pattern was observed by aligning one arm of the interferometer parallel to the direction of the rotation of earth, and the other perpendicular to it. Next the interferometer was rotated by 90°, and if ether existed a shift of interference pattern by about 0.4 fringe using light of wavelength of ~ 5000 $\mathring{A}$ was expected. The experiment was performed at all hours of the day, in all positions of the earth in its orbit, yet no fringe shift was ever observed, although the experimental setup was capable of detecting a shift of upto 0.01 fringe. This once for all negated the existance of ether, that set the Modern Physics on a new course.

6.7.5 Fabry-Perot Interferometer

Fabry-Perot Interferometer is a simpler, less expensive version of Michelson interferometer, that consists of two glass plates P_1 and P_2, partly silvered on one side such that the reflectivity is $80 - 90\%$ The plates are held parallel to each other at a separation of 0.1 to 10 cm with silvered surfaces facing each other such that an air film is enclosed between them. One of the plate is fixed while the other can be moved parallel to itself, changing the thickness of the air film.(Figure 6.19).

When the plates assembly is illuminated by monochromatic light from a wide source, an interference pattern of concentric circular fringes is observed in the light transmitted from the other side of the plates assembly. The condition of constructive interference is:

$$2d\,cos\theta = m\lambda \qquad (Maxima) \tag{6.71}$$

where we have taken $n_{air} = 1$, so that the angle of refraction $\phi' = \theta$, the angle of incidence. On varying the distance between the plates, the interference fringes appear to emerge from or sink into the center in the same way as in Michelson interferometer. From the number of interference fringes disappearing or appearing for a displacement Δd of the plate and taking $\theta = 0$ for the center of the fringes, the wavelength of light can be determined. If the plates of the Febry-Perot interferometer are mounted at a fixed separation, it is called Febry-Perot etalon, and is used for various spectroscopic applications.

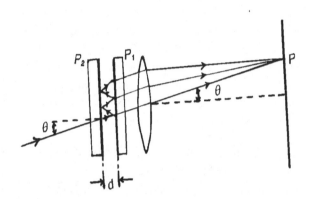

Figure 6.19: Interference by Febry-Perot interferometer.

6.8 Chapter 6 Summary

This Chapter deals with Interference of Light. It covers superposition of waves, requirements to observe interference of light and conditions of interference, phasor addition of waves, interference by division of wavefront, and by division of amplitude.

- Principle of superposition of waves: When two or more waves travel simultaneously though a medium/ space, the resultant displacement of medium/ in space is the vector addition of the displacement due to individual waves.

- Simplest example is the simultaneous propagation of two waves in water. Displacements due to two waves in the same direction add to give a large displacement of water (constructive interference) and displacements in opposite directions result in a smaller displacement of water or zero displacement if both waves are equal (destructive interference).

- Because of the small wavelength of light, to observe a stable interference pattern the two sources of light must be monochromatic (single wavelength) and coherent, (constant phase relationship between the two waves). If white light is used, first few overlapping coloured interference fringes are observed after which the pattern becomes hazy mix of colours and beyond that uniform white illumination is obtained.

- For most applications sodium bulb, although it produces two close wavelengths with ??$= 0.597$ nm, is an adequate source of monochromatic light. one can also produce monochromatic light spectroscopically or may use lasers.

- Pair of coherent sources is produced either by division of wavefront or by division of amplitude from the same source of monochromatic light.

- Amplitude and phase of the resultant wave motion from the superposition of a number of simultaneous waves can be obtained graphically by the vector addition of the amplitude and phase of individual wave in the same manner as the graphical addition of magnitude and direction of vectors (Re: Vector algebra).

- optical path difference and phase difference: ○ optical path difference between two waves travelling simultaneously in a medium of refractive index n is: $\Delta x =$ (Physical path difference) x_n. ○ Phase difference between the same pair of waves is: $\Delta\phi = (2\pi/\lambda) \times \Delta x$. ○ A wave on reflection from a surface backed by a denser medium (glass, mirror, water) undergoes a phase change of π. This results in additional path difference of $(\lambda/2)$ and a phase difference of π if one of the coherent wave is obtained by reflection and the other is not.

- Conditions of constructive and destructive interference for two waves, none of which undergoes reflection or if both undergo reflection before interference: ○ Two waves interfere constructively if their optical path difference at the point of superposition is an integer multiple of $\lambda(\Delta x = m\lambda, m = 0, 1, 2, \cdots)$ or if the phase difference is integer multiple of $2\pi(\Delta\phi = 2m\pi, m = 0, 1, 2, \cdots)$. ○ Two waves interfere destructively if their optical path difference at the point of superposition is odd-integer multiple of $(\lambda/2)(\Delta x = (2m+1)(\lambda/2), m = 0, 1, 2, \cdots)$ or if the phase difference is odd-integer multiple of $\pi(\Delta\phi = (2m+1)\pi, m = 0, 1, 2, \cdots)$.

- If one of the waves undergoes reflection before interference, because of an additional phase of π for the reflected wave, the conditions of constructive and destructive interference given above are interchanged.

- Interference by division of wave front: Wave front from a source slit is split into two portions, resulting into two coherent sources both of which may be real or both virtual or one real and one virtual. Interference can only be observed under controlled laboratory conditions. This has no practical/ industrial application except measuring of unknown wavelength of light. The following experimental techniques and the resulting interference pattern are presented.
○ Youngs double slit experiment: Both coherent sources are real, fringes are of equal intensity, central fringe is bright and their separation depends on d and D.
○ Fresnel biprism: Both the sources are virtual, and the interference pattern is similar to that from double slits.
○ Fresnels mirrors: Both the sources are virtual, both the split wavefronts undergo a phase change of p on reflection which nullify each other, and hence the interference pattern is similar

to the double slits pattern.

○ Lloyds mirror: one of the beam is obtained by reflection from a mirror which undergoes phase change of π' and the corresponding source is virtual. The other beam is obtained directly from the source which is real. The conditions of interference are reversed.

- Interference by division of amplitude: Intensity of the incident wave and hence the amplitude is split in to two interfering beams by reflection from the front surface, and refraction and subsequent reflection from the back surface. Interference can also be observed in the transmitted beam. Interference phenomenon is widely and freely observable in nature, such as the colours of thin oil films on water and soap bubbles in white light, and have practical/ technological applications. The following experimental techniques and interference pattern produced by them are presented.

○ Thin parallel film: This has application in anti-reflection coating of optical components and solar cells. ○ Wedge shaped film is produced by separating one end of two glass slides held together with a thin spacer like paper or wire. By introducing a drop of a liquid in the wedge refractive index of the liquid can be measured. ○ Newtons rings: Again used for the measurement of unknown wavelengths, refractive index of rare/ small quantity of liquids, difference of two close wavelengths for example the two spectral lines of sodium lamp. ○ Michelson interferometer: Historically this was designed to negate the presence of Ether and has applications in measurement of wavelength, small wavelength difference, refractive index, and standardization of the meter in terms of the wavelength of light. It is an essential teaching aid in the college optics laboratories. ○ Fabry-Perot interferometer: It is the compact version of the Michelson interferometer, works on the same principle and has same applications.

6.9 Exercises

6.1 What are the essential differences between sound and light waves? In what respect are they similar?

6.2 Three wave motions given by:
$$y_1 = 3sin(\omega t)$$

$$y_2 = 5sin(\omega t + \frac{\pi}{3})$$

and
$$y_3 = 7sin(\omega t + \frac{2\pi}{3})$$

simultaneously travel through a medium. *(i)* On one single diagram plot the phasor diagrams of all the wave motions. *(ii)* Find the amplitude and phase of the resultant wave motion, and *(iii)* Write down the equation of the resultant wave motion.

6.3 In exercise 6.2, consider that all the three wave motions have equal amplitude of 4 units, and find the resultant wave motion graphically, and verify your result analytically.

6.4 Ten waves of the same frequency and amplitude, and with successive increasing phase difference $\phi, 2\phi, 3\phi, 4\phi, \ldots$ are superimposed to produce an interference. *(i)* Determine the value of ϕ if the resultant wave is a minimum of interference of zero intensity. *(ii)* What is the non-zero value of ϕ for a maxima of the largest possible intensity.

6.5 In a Young's double slit experiment separation between the screen and slit is kept constant, and the separation of double slits is adjustable. In four separate experiments interference patterns are observed using lights of the four wavelengths given in Table 6.1, and the slits separation is adjusted such that the fringe separation in each case remains constant. Calculate the ratio of slit separation in each case.

6.6 In exercise 6.5, keep the slits separation constant, and change the separation between the slits and the screen such that the interference fringes separation in each case remains constant. Calculate the ratio of the screen separation.

6.7 Describe the Young's double slit interference pattern observed using white light. In an experimental set up separation between the slits and the screen is 1 m, and the separation between the slits is 0.25 mm. When one of the slits is covered with a 1 mm thick plate of a transparent plastic material, the central interference fringe with white light is displaced by 5 mm on the screen. If the average wavelength of light is 550 nm, calculate the refractive index of the plastic.

6.8 In a Young's double slit experiment interference pattern is observed using sodium light. The slits separation is 0.25 mm, and the observation screen is placed at a distance of 1m from the slits. Calculate the separation of the tenth bright fringe on either side of the central fringe due to both the wavelengths in sodium light. Repeat calculations for light of wavelengths given in table 6.1.

6.9 In a Fresnel biprism experiment, the screen is placed 1 m away from the slit, and the separation between successive bright fringes is measured to be 0.1002 cm. A convex lens is used to measure the separation between the two virtual coherent sources. For two possible locations of the lens between the screen and the biprism, images of the coherent sources are formed on the screen which are 2.86 mm and 4.13 mm apart. Calculate *(i)* the separation between the pair of the coherent sources, and *(ii)* wavelength of light.

6.10 Describe the biprism interference produced with white light. When one of the prisms of the biprism is covered with a thin mica film of refractive index 1.60, the central fringe of white light shifts by 2.05 mm. If the average wavelength of white light is taken to be 590 nm, calculate the thickness of the mica plate. How the interference pattern shall change if the same mica plate is placed in front of the source slit?

6.11 In Young's double slit experiment, the separation for two slits is 0.02 mm and the screen is placed at 80 cm from the slits. Using light of wavelength λ separation of successive bright fringes is measured. The experiment is now repeated using Fresnel's mirrors with an angle of 2° between them. *(i)* How far should the mirrors be placed so that the separation between the pair of virtual sources is the same as the separation of the Young's double slits. *(ii)* How far from the mirrors should the screen be placed so that the fringe separation is twice the fringe separation in the double slit experiment?

6.12 In a Lloyd mirror experiment the source of sodium light is placed 1 mm above the mirror surface, and the mirror is 20 cm long. Find the separation of two successive bright fringes if the screen is placed *(i)* at the far edge of the mirror, *(ii)* 70 cm from the far edge of the mirror. What wavelength shall produce *(iii)* the same fringe separation at 70 cm as that of sodium light fringes at the edge of the mirror, *(iv)* the same fringe separation at the edge of the mirror as that of sodium light fringes at 70 cm location of the screen? Do these wavelengths lie in the visible region of spectra?

6.13 Interference patterns of red and blue light are studied using Lloyd's mirror of length *L*. The slit is at a height *d* from the mirror surface, and the screen is placed at the far edge of the mirror. *(i)* Which wavelength has a larger separation of fringes? *(ii)* How far should the screen be moved so that the smaller fringes separation becomes equal to the larger fringe separation when screen was placed at the edge of the mirror? *(iii)* What shall be the fringe separation of the other wavelength at this new location of the screen? *(iv)* What should be the height of the slit in units of *d* so that the fringe separation of blue light is the same as that of the red light for a fixed location of the screen? *(v)* What should be the height of the slit in units of *d* so that the fringe separation of red light is the same as that of the blue light for a fixed location of the screen?

6.14 A plane parallel film of thickness 0.0035 cm has a refractive index of 1.35. It is illuminated with sodium light at $\theta = 0^o, 30^o, 45^o, 60^o$, and 75° Calculate the order of interference in each case, and state if the interference is constructive or destructive. For the same film, calculate the angles of incident for the first four maxima, and the first four minima in the interference with red light ($\lambda_R = 700nm$), assuming the refractive index to be the same as given above.

6.15 A thin parallel soap film ($n = 1.33$) in sodium light at normal incident appears dark. Calculate *(i)* the minimum thickness of the film, *(ii)* the minimum angle of incident at which it will appear bright. The thickess of the film slowly decreases due to evaporation of water while the film remains plane parallel. Assume that the refractive index also remains constant. Then calculate *(iii)* at what thickness the film shall appear bright on normal incident, and *(iv)* at what angle of incident it shall appear dark?

6.16 Calculate the minimum thickness and refractive index of the anti-reflection coating for a glass filter ($n = 1.563$) for green light $\lambda = 550nm$.

6.17 A soap film is formed on a rectangular wire frame that is held vertically. Describe the shape of the film, and the interference pattern that shall be seen in the reflected light from this film. Using light of wavelength 590 nm, separation between two consequtive bright fringes is found to be 5 mm. If refractive index of the film is 1.33, calculate the angle of the film.

6.18 The same soap film as in exercise 6.17 is viewed in reflected white light. Calculate the distances of the 10th red and blue bright fringes from the edge of the film. What other wavelengths in the visible region will produce constructive interference at these locations, and what wavelengths shall produce destructive interference? This exercise demonstrates how various order maximum and minimum due to different wavelengths overlap when white light is used.

6.19 In Newton's rings experiment with light of wavelength 590 nm, the diameters of two consecutive rings in reflected light are measured to be 2.00 cm and 2.02 cm. Calculate the radius of curvature of the lens. Now a drop of water ($n = 1.33$) is introduced between the lens and the glass plate. Calculate the diameters of the same rings. What orders of rings with water are observed that have the same diameters *i.e.* 2.00 cm and 2.02 cm respectively?

6.20 Newton's rings are observed in reflected sodium light by placing one convex lens of radius of curvature 10 cm on another convex lens of radius of curvature 20 cm. Calculate the diameter of the twentieth bright ring. If a drop of water is introduced between the lenses, what shall be the diameter of the twentieth ring, and what order interference ring shall be observed at the same diameter as the twentieth ring without water?

6.21 In Michelson interferometer 130 fringes sink into the center when the movable mirror is moved a distance of 0.04 mm. Calculate the wavelength of light. With the same light, how far, and in which direction relative to the previous motion should the mirror be moved for 200 fringes to emerge from the center of the interference pattern?

6.22 Using light of wavelength 600 nm, interference pattern of circular fringes is produced with a tube of 5 cm length in the path of one of the beams in Michelson interferometer. Now the tube is slowly evacuated while counting the interference fringes that shift. If 49 fringes have shifted with the full evacuation of the tube, calculate the refractive index of air. Now the tube is slowly filled with another gas, and 250 fringes are observed to shift. What is the refractive index of the gas? By how much should the mirror be moved to restore the interference pattern to exactly to the same state that was observed with the air in the tube?

6.23 If you were to calibrate a meter in terms of the two spectral lines of sodium, what shall be the meter standard in terms of these wavelengths. How many points of consonance and dissonance shall be observed when the movable mirror is moved one full meter length?

6.24 A Febry-Perot interferometer is used to determine the difference of two close wavelengths, the shorter one of which is 546.074 nm. If the consonance of fringes due to two wavelengths occur at the separations of the plates equal to 0.649 mm, 1.829 mm, and 3.009 mm, calculate the longer wavelength.

Chapter 7

Diffraction of Light

7.1   Introduction to Diffraction: Fraunhofer and Fresnel Diffraction

The observation that light travels in straight lines is well known and was discussed in chapter 1 under the principle of rectilinear propagation of light. Rectilinear propagation of light leads to the observation of shadows cast when light illuminates objects. However, there is another observation whereby when light passes through some obstacles it appears to cast not an absolute dark shadow, but rather a pattern with dark and light bands. This is what is referred to as *diffraction*, and this chapter is devoted to studying this effect. Diffraction is a property that demonstrates the wave aspect. Diffraction occurs when the dimensions of the obstacle interacting with a wave is of the same order of magnitude as the wavelength.

Diffraction effects are conveniently classified into *Fraunhofer diffraction* and *Fresnel diffraction*. Fraunhofer diffraction occurs when the light source and the screen where the diffraction pattern is observed are effectively at infinite distances from the aperture or obstacle causing diffraction. Fresnel diffraction is such that either the source of light or the screen or both are at finite distances from the aperture causing diffraction.

In this chapter, we shall consider several examples of Fraunhofer diffraction, namely

- Single slit diffraction

- Rectangular aperture diffraction

- Circular aperture diffraction

- Double slits diffraction (Young's slits revisited)

- N-slit diffraction

Also, we discuss Fresnel diffraction by straight edges. The chapter concludes with a discussion of X-ray diffraction, electron diffraction and neutron diffraction.

7.2   Single slit diffraction

Consider light of angular frequency $\omega(= 2\pi\nu$, with ν being frequency) and wavevector $k(= 2\pi/\lambda$, with λ being the wavelength) incident on a single slit S_1 of width d as illustrated in Figure 7.1. The light is considered to be from a distant source, and can therefore be regarded as a plane wave. We wish to calculate the intensity distribution observed on a screen S_2 at a distance D away.

The field of a wave emitted by a secondary source centred at O is given by

$$Re\left(A_0 e^{i(kr-\omega t)}\right) \tag{7.1}$$

where Re represents the real part of the field.

Consider an element of length dx at a distance x from O. The field of a wave emitted by a secondary source centred at an element dx is given by

$$Re A_0 e^{i(kr_1-\omega t)} = Re A_0 e^{i(kr-\omega t)} e^{-ikx\sin\theta} \tag{7.2}$$

where from Figure 7.1, we have used $r = r_1 + x\sin\theta$.

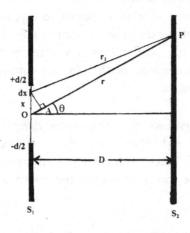

Figure 7.1: An illustration of Fraunhofer diffraction by a single slit of width d.

The total observable field will be the sum of all the elements from $-d/2$ to $+d/2$, given by $A(r,\theta)$

$$
\begin{aligned}
A(r,\theta) &= Re\int_{-d/2}^{+d/2} A_0 e^{i(kr-\omega t)} e^{-ikx\sin\theta} dx \\
&= A_0 d\frac{\sin\{\frac{1}{2}kd\sin\theta\}}{\{\frac{1}{2}kd\sin\theta\}}\cos(kr-\omega t) \\
&= A(0)\frac{\sin\beta}{\beta}\cos(kr-\omega t) \\
&= A(\theta)\cos(kr-\omega t)
\end{aligned}
\tag{7.3}
$$

where $A(0) = A_0 d$, and the amplitude $A(\theta)$ of the resultant field is given by

$$A(\theta) = A(0)\frac{\sin \beta}{\beta} \tag{7.4}$$

$$\beta = \frac{1}{2}kd \sin \theta \tag{7.5}$$

The intensity distribution is the square of the amplitude distribution, given by

$$I(\theta) = I(0)\frac{\sin^2 \beta}{\beta^2} \tag{7.6}$$

where $I(0) = A^2(0)$. The intensity distribution given in equation 7.6 has a series of bright (maxima) and dark (minima) fringes. The graphical illustration of $I(\theta)$ against β is shown in Figure 7.2.

The function $\sin \beta / \beta$ is sometimes referred to as $sinc\beta$, in terms of the $sinc$ function.

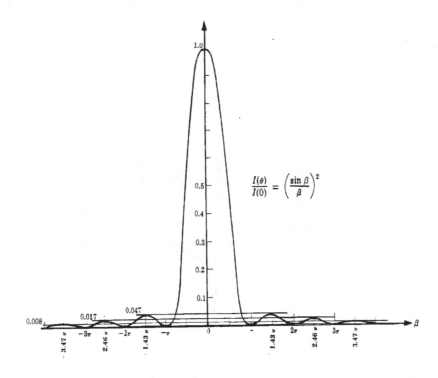

Figure 7.2: A graph of $I(\theta)/I(0)$ against β for single slit Fraunhofer diffraction.

The maxima and minima in the graph of $I(\theta)$ against β can be found by differentiating the expression for $I(\theta)$ with respect to β, as shown below.

$$\frac{dI}{d\beta} = \frac{d}{d\beta}\{I(0)\frac{\sin^2 \beta}{\beta^2}\}$$

$$= I(0)\frac{2\sin\beta(\beta\cos\beta - \sin\beta)}{\beta^3}$$

$$= 0 \tag{7.7}$$

which implies that minima (or dark fringes) occur when $\sin\beta = 0$ and $\beta \neq 0$, which is when

$$\beta = \pm n\pi \text{ when n=1,2,3,} \cdots \tag{7.8}$$

The maximum central peak occurs at $\beta = 0$. The subsidiary maxima exist for nonzero β satisfying

$$\beta\cos\beta - \sin\beta = 0 \tag{7.9}$$

or

$$\tan\beta = \beta \tag{7.10}$$

Equation (7.10) can be solved graphically as illustrated in Figure 7.3 by finding the points of intersection of two functions, f_1 and f_2, with $f_1 = \beta$ and $f_2 = \tan\beta$, and the values of β are found as

$$\beta = \pm 1.4303\pi, \pm 2.4590\pi, \pm 3.4707\pi, \cdots.$$

The secondary maxima are not quite midway the minima points.

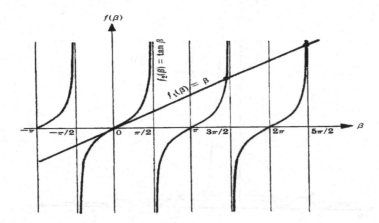

Figure 7.3: Graphs of $f_1 = \beta$ against β and $f_2 = \tan\beta$ against β. Solutions of the equation $\tan\beta = \beta$ are the points of intersection of the two functions, f_1 and f_2.

Diffraction affects the ability of optical instruments, such as microscopes and telescopes to distinguish between closely spaced objects. This is what is referred to as *resolution* of images. To understand this, consider equation 7.5 and 7.8 from single slit diffraction. The equations imply

$$\sin\theta = n\frac{\lambda}{d}$$

or, considering the central bright fringe ($n = 1$), and noting that for small values of θ, the approximation $\sin\theta \approx \theta$ can be used, one obtains

$$\theta = \frac{\lambda}{d}$$

where "θ" is known as the *half angular width*. "θ" sets the limit of resolution of an optical instrument. For example, a question arises, when are two images resolved? This is answered by considering what is referred to as *Rayleigh's criterion..*

Rayleigh's criterion states that two images or point sources are just resolved when the central maxima of one just coincides with the first minima of the diffraction of the other. This in turn implies that the central maxima due to the two sources must be separated by at least half the angular width "θ" of the central maxima.

7.3 Diffraction by a rectangular aperture

Consider light of angular frequency $\omega(= 2\pi\nu$, with ν being frequency) and wavevector $k(= 2\pi/\lambda$, with λ being the wavelength) incident from far on a rectangular aperture of width a and height b as illustrated in Figure 7.4.

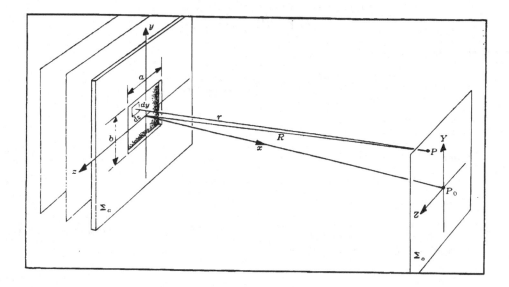

Figure 7.4: An illustration of Fraunhofer diffraction by a rectangular aperture of width a and height b.

We wish to calculate the intensity distribution observed on a screen S_2 at a distance x away. Each element of area $dS = dydz$ is a source of wavelets. An element at $(0, y, z)$ produces a field dE_P at point $P(x, Y, Z)$ on screen S_2, of the form

$$dE_P = \frac{E_0}{r} e^{i(kr-\omega t)} \tag{7.11}$$

where from figure 7.4:

$$r = [x^2 + (Y - y)^2 + (Z - z)^2]^{1/2} \tag{7.12}$$

$$R = [x^2 + Y^2 + Z^2]^{1/2} \tag{7.13}$$

In view of R being large, $y^2 \ll 2Yy$ and $z^2 \ll 2Zz$, the following approximation can be introduced

$$r \approx R\{1 - \frac{Yy + Zz}{R^2}\} \tag{7.14}$$

The total observable field will be the sum of all the elements from $-b/2$ to $+b/2$ along y and from $-a/2$ to $+a/2$ along z, given by the real part of E_P.

$$\begin{aligned}
E_P &= \frac{E_0}{R} e^{i(kR-\omega t)} \int_{-b/2}^{+b/2} \int_{-a/2}^{+a/2} e^{-ik(Yy+Zz)/R} dydz \tag{7.15}\\
&= \frac{E_0}{R} e^{i(kR-\omega t)} \int_{-b/2}^{+b/2} e^{-ikYy/R} dy \int_{-a/2}^{+a/2} e^{-ikZz/R} dz \\
&= \frac{abE_0}{R} e^{i(kR-\omega t)} \frac{\sin(kbY/2R)}{(kbY/2R)} \frac{\sin(kaZ/2R)}{(kaZ/2R)} \tag{7.16}
\end{aligned}$$

Taking the real part of E_P,

$$\begin{aligned}
Re\,(E_P) &= \frac{abE_0}{R} \frac{\sin(kbY/2R)}{(kbY/2R)} \frac{\sin(kaZ/2R)}{(kaZ/2R)} \cos(kR - \omega t) \\
&= \frac{abE_0}{R} \frac{\sin\beta}{\beta} \frac{\sin\alpha}{\alpha} \cos(kR - \omega t) \tag{7.17}\\
\text{where } \beta &= kbY/2R \tag{7.18}\\
\alpha &= kaZ/2R \tag{7.19}\\
Re\,(E_P) &= E(Y, Z)\cos(kR - \omega t) \tag{7.20}
\end{aligned}$$

where $E(Y, Z)$ is the amplitude. The intensity distribution due to diffraction by a rectangular aperture is the square of the amplitude distribution, given by

$$I(Y, Z) = E^2(Y, Z) \tag{7.21}$$

or

$$I(Y, Z) = I(0) \frac{\sin^2\beta}{\beta^2} \frac{\sin^2\alpha}{\alpha^2} \tag{7.22}$$

a result which could have been anticipated from the result of diffraction by a single slit given in equation (7.6). The photographic illustration of $I(Y, Z)$ is shown in Figure 7.5.

Figure 7.5: A photograph of the diffraction pattern produced by a rectangular aperture.

7.4 Diffraction by a circular aperture

Consider light of angular frequency $\omega(= 2\pi\nu$, with ν being frequency) and wavevector $k(= 2\pi/\lambda$, with λ being the wavelength) incident from far on a circular aperture of diameter d as illustrated in Figure 7.6.

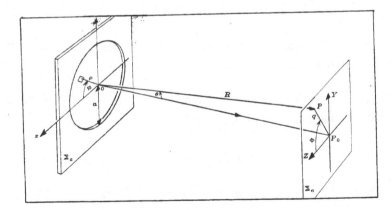

Figure 7.6: An illustration of Fraunhofer diffraction by a circular aperture of diameter d.

Without going into mathematical details, the intensity distribution observed on a screen at a distance away from the aperture, is given by

$$I(\theta) \;=\; I(0)\left[\frac{2J_1(\tfrac{1}{2}kd\sin\theta)}{\tfrac{1}{2}kd\sin\theta}\right]^2$$

$$\qquad\;=\; I(0)\left[\frac{2J_1(u)}{u}\right]^2 \tag{7.23}$$

$$\text{where } u \;=\; \frac{1}{2}kd\sin\theta \tag{7.24}$$

and $J_1(u)$ is the first order Bessel function defined by the series

$$J_1(u) = \frac{u}{2}\left[1 - \frac{1}{1!2!}\left(\frac{u}{2}\right)^2 + \frac{1}{2!3!}\left(\frac{u}{2}\right)^4 - \frac{1}{3!4!}\left(\frac{u}{2}\right)^6 + \cdots\right] \tag{7.25}$$

The intensity distribution of the diffraction pattern due to a circular aperture shows a series of maxima and minima as illustrated in the graph of $I(\theta)/I(0)$ against $\tfrac{1}{2}kd\sin\theta$ in Figure 7.7. A photograph of the diffraction pattern produced by a circular aperture is shown in Figure 7.8, where it can be noted that there is a central bright disc surrounded by alternating dark and bright rings, with progressively fainter rings.

The maxima of intensity occur at

$$\frac{1}{2}kd\sin\theta = 0, \pm 5.14, \pm 8.42, \cdots$$

and the minima of intensity occur at

$$\frac{1}{2}kd\sin\theta = 3.83, \pm 7.02, \cdots$$

As in the case of single slit diffraction, the dominant maxima is at $\theta = 0$, and the secondary maxima are not halfway between the minima.

An application of the concept of *resolution* of images discussed in section 7.2 to diffraction by a circular aperture shows that the half angular width is

$$\theta = 1.22\frac{\lambda}{d}$$

where d is the diameter of the circular aperture and the factor of 1.22 arises from the Bessel function mentioned earlier, in equation 7.23.

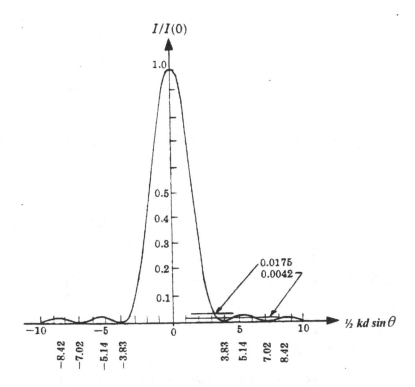

Figure 7.7: A graph of $I(\theta)/I(0)$ against $\frac{1}{2}kd\sin\theta$ for Fraunhofer diffraction by a circular aperture.

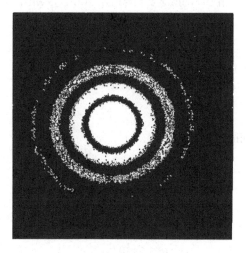

Figure 7.8: A photograph of the diffraction pattern produced by a circular aperture.

7.5   Double slit diffraction (Young's slits)

Consider light of angular frequency $\omega(= 2\pi\nu$, with ν being frequency) and wavevector $k(= 2\pi/\lambda$, with λ being the wavelength) incident from far on a system with double-slits (also known as Young's slits) , S_1 and S_2, with a slit-spacing d as illustrated in Figure 7.9. We wish to calculate the intensity distribution observed at a point P on a screen at a distance D from the double slits.

It should be noted that we are revisiting the system of Young's slits which was initially discussed in chapter 6 on interference. The reason why we are studying this system again is to emphasize that each of the two slits produces diffraction effects as was discussed in section 7.2 when we considered single slit diffraction. Subsequently, the two diffracted intensities will superimpose to produce the resulting interference pattern.

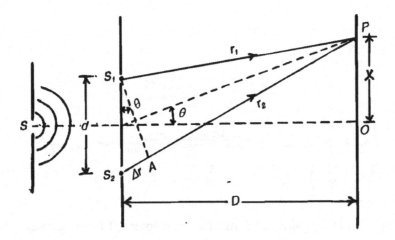

Figure 7.9: An illustration of Young's slits, S_1 and S_2, with a slit-spacing d.

The following observations can be made.
From the geometry, it can be noted that $D \gg d, D \gg x$ and $d \gg \lambda$, and
$r_2 = r_1 + d\sin\theta$
$2r_2 = (r_1 + r_2) + d\sin\theta$
An average distance, r, can be defined as

$$r = \frac{1}{2}(r_1 + r_2)$$

and hence $r_2 = r + \frac{1}{2}d\sin\theta$ and $r_1 = r - \frac{1}{2}d\sin\theta$
The path difference between waves through S_1 and S_2 is AS_2 given by

$$AS_2 = d\sin\theta$$

There is another quantity of physical interest, the phase difference, ϕ, given by

$$\phi = \frac{2\pi}{\lambda} \times \text{Path difference}$$

$$
\begin{aligned}
&= \frac{2\pi}{\lambda} d \sin\theta \\
&= kd \sin\theta
\end{aligned}
\tag{7.26}
$$

The total observable field at P will be the sum of the fields through *each* of the double-slits, given by y_P in the form

$$
y_P = a e^{i(kr_1 - \omega t)} + a e^{i(kr_2 - \omega t)}
\tag{7.27}
$$

where a is the diffracted amplitude throught each of the slits, given by equation 7.4, that is

$$
a = A(0) \frac{\sin\beta}{\beta}
$$

and $\beta = \frac{1}{2} kb \sin\theta$, where we are introducing a slit width b for each of the slits, which is different from the slit separation d shown in Figure 7.9. We can express kr_1 and kr_2 in terms of ϕ as:

$$
\begin{aligned}
kr_2 &= kr + \frac{\phi}{2} \\
kr_1 &= kr - \frac{\phi}{2}, \text{ and hence} \\
y_P &= a e^{i(kr - \omega t - \phi/2)} + a e^{i(kr - \omega t + \phi/2)} \\
&= a e^{i(kr - \omega t)} \{ e^{i\phi/2} + e^{-i\phi/2} \} \\
&= 2A(0) \frac{\sin\beta}{\beta} \cos\frac{\phi}{2} e^{i(kr - \omega t)}
\end{aligned}
\tag{7.28}
$$

Taking the real part, we obtain $Re\ y_P$,

$$
\begin{aligned}
Re\ y_P &= Re\ 2A(0) \frac{\sin\beta}{\beta} \cos\frac{\phi}{2} e^{i(kr - \omega t)} \\
&= 2A(0) \frac{\sin\beta}{\beta} \cos\frac{\phi}{2} \cos(kr - \omega t) \\
&= A(r, \theta) \cos(kr - \omega t)
\end{aligned}
\tag{7.29}
$$

where

$$
\begin{aligned}
A(r, \theta) &= 2A(0) \frac{\sin\beta}{\beta} \cos\left(\frac{\phi}{2}\right) \\
&= 2A(0) \frac{\sin\beta}{\beta} \cos\gamma
\end{aligned}
\tag{7.30}
$$

where we have introduced $\gamma = \phi/2$. The net intensity distribution observed on the screen is the square of the amplitude, given by

$$
\begin{aligned}
I(\theta) &= A^2(r, \theta) \\
&= 4A^2(0) \frac{\sin^2\beta}{\beta^2} \cos^2\frac{\phi}{2} \\
&= 4A^2(0) \frac{\sin^2\beta}{\beta^2} \cos^2\gamma
\end{aligned}
\tag{7.31}
$$

where one can identify

$$\frac{\sin^2 \beta}{\beta^2} \qquad (7.32)$$

as the *diffraction term*, with $\beta = \frac{1}{2}kb\sin\theta$, and

$$\cos^2 \gamma \qquad (7.33)$$

as the *interference term*, with $\gamma = \phi/2 = (1/2)kd\sin\theta$.

The resultant intensity will have a series of maxima (bright fringes) and minima (dark fringes). This can be understood as follows. The resultant intensity given in equation 7.31 will be zero if either the diffraction term is zero or the interference term is zero

The diffraction term is zero (*minima*) when

$$\beta = \pm n\pi \text{ when n=1,2,3,}\cdots$$

which implies

$$\beta = \frac{1}{2}kb\sin\theta = \pm n\pi \text{ when n=1,2,3,}\cdots$$

or

$$b\sin\theta = \pm n\lambda = \pm\lambda, \pm 2\lambda, \pm 3\lambda, \cdots$$

The interference term is zero (*minima*) when

$$\gamma = (m + \frac{1}{2})\pi \text{ where } m = 0, 1, 2, \cdots$$

that is, $(m + \frac{1}{2})$ is half odd integer, and thus

$$\frac{1}{2}kd\sin\theta = \frac{\pi}{\lambda}d\sin\theta = (m + \frac{1}{2})\pi$$

or

$$d\sin\theta = (m + \frac{1}{2})\lambda = \frac{1}{2}\lambda, \frac{3}{2}\lambda, \frac{5}{2}\lambda, \cdots$$

The position of the *maxima* are determined by $\beta = 0$ and also

$$\gamma = m\pi \text{ where } m = 0, 1, 2, \cdots$$

that is, m is an integer, and thus

$$\frac{1}{2}kd\sin\theta = \frac{\pi}{\lambda}d\sin\theta = m\pi$$

or

$$d\sin\theta = m\lambda = 0, \lambda, 2\lambda, \cdots$$

A graph of the diffraction term is of illustrated in Figure 7.10 (a), while that of the interference term is illustrated in Figure 7.10(b). The products of these two graphs gives the diffracted intensity through the Young's slits, as illustrated in Figure 7.10(c).

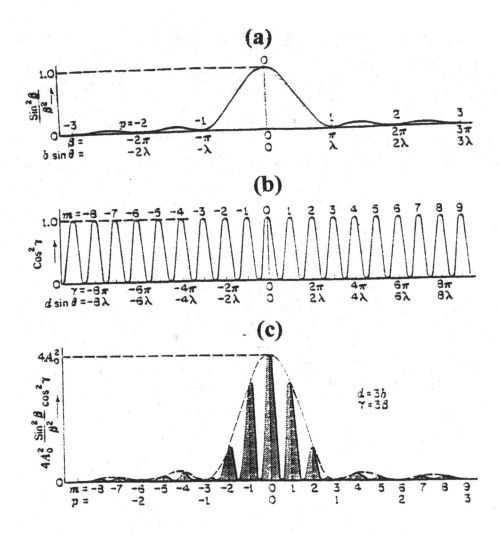

Figure 7.10: The intensity distribution of light through Young's slits, showing (a) the diffraction term (b) the interference term (c) the product of the diffraction term and the interference term which gives the resultant intensity distribution.

Form Figure 7.10 one notes the following interesting features of the double slit diffraction pattern.

- The spacing between the maxima and minima of the fringes is determined by the interference effect, whereas the intensities of maxima are determined by the diffraction effect. Thus, while the fringes are equally spaced, they are not equally bright. The central fringe is the brightest, and the intensity diminishes on either side of it.

- The diffraction pattern of the single slit acts as the envelope of the interference fringes. As a result of the minima of the diffraction pattern some of the maxima of the interference pattern are suppressed. These are called the missing orders, and they occur where the condition of diffraction minimum coincides with the condition of the interference maximum.

7.6   N-slits diffraction or Diffraction Grating

Consider light of angular frequency $\omega (= 2\pi\nu$, with ν being frequency) and wavevector $k(= 2\pi/\lambda$, with λ being the wavelength) incident from far on a system with N-slits, each of width b and the slit-spacing is d as illustrated in Figure 7.11. This system is referred to as a diffraction grating.

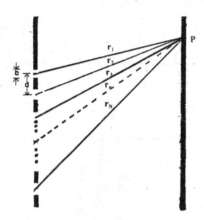

Figure 7.11:A diffraction grating with N-slits, each of width b and slit-spacing d.

We wish to calculate the intensity distribution observed at a point P. The following observations can be made.

First, there will be single slit diffraction from each of the N-slits. *Second*, there will be interference of all the diffracted beams arising from each of the N-slits.

From geometry,
$r_2 - r_1 = d\sin\theta$.
$r_3 - r_1 = 2d\sin\theta$, and generally, we have
$r_n - r_1 = (n-1)d\sin\theta$.
An average distance, r_{av}, can be defined as

$$r_{av} = \frac{1}{2}(r_1 + r_N) = r_1 + (N-1)\frac{d}{2}\sin\theta$$

The total observable amplitude at P will be the sum of all diffracted amplitudes through the N-slits, given by E_P in the form

$$E_P = Ae^{-i\omega t}\{e^{ikr_1} + e^{ikr_2} + \cdots + e^{ikr_N}\} \qquad (7.34)$$

where each of the slits contributes the amplitude A, and as discussed in the section dealing with single slit diffraction $A = A(0)\frac{\sin\beta}{\beta}$, as was shown in equation 7.4.
But note that

$$
\begin{aligned}
e^{ikr_n} &= e^{i[kr_1 + (n-1)kd\sin\theta]} \\
&= e^{i[kr_1 + (n-1)\phi]} \\
&= e^{ikr_1}e^{i(n-1)\phi} \qquad (7.35) \\
\text{where } \phi &= kd\sin\theta \qquad (7.36)
\end{aligned}
$$

and using this in equation (7.34), the total observable field at P can be written in the form

$$
\begin{aligned}
E_P &= Ae^{-i\omega t}e^{ikr_1}\{1 + e^{i\phi} + e^{2i\phi} + \cdots + e^{i(N-1)\phi}\} \\
&= Ae^{-i\omega t}e^{ikr_1}\{1 + a + a^2 + \cdots + a^{(N-1)}\} \qquad (7.37) \\
\text{where } a &= e^{i\phi} \qquad (7.38)
\end{aligned}
$$

Note that the terms in the curly brackets in equation (7.37) form a geometrical progression (GP) of N terms, with the first term in the curly brackets as 1, and the common ratio is $e^{i\phi}$, and thus the sum is given by

$$
\begin{aligned}
S_{GP}^N &= \frac{[a^N - 1]}{[a - 1]} \\
&= \frac{[e^{iN\phi} - 1]}{[e^{i\phi} - 1]} \\
&= \frac{e^{iN\phi/2}[e^{iN\phi/2} - e^{-iN\phi/2}]}{e^{i\phi/2}[e^{i\phi/2} - e^{-i\phi/2}]} \\
&= e^{i(N-1)\phi/2}\frac{\sin N\phi/2}{\sin\phi/2} \qquad (7.39)
\end{aligned}
$$

Using these results in equation (7.37), the total observable field at P can be written in the form

$$
\begin{aligned}
E_P &= Ae^{-i\omega t}e^{[ikr_1 + i(N-1)\phi/2]}\frac{\sin N\phi/2}{\sin\phi/2} \\
&= Ae^{i(kr_{av}-\omega t)}\frac{\sin N\phi/2}{\sin\phi/2} \qquad (7.40)
\end{aligned}
$$

Taking the real part, we obtain $ReE_P = E(r,\theta)$,

$$
\begin{aligned}
ReE_P &= ReAe^{i(kr_{av}-\omega t)}\frac{\sin N\phi/2}{\sin\phi/2} \\
E(r,\theta) &= A\cos(kr_{av} - \omega t)\frac{\sin N\phi/2}{\sin\phi/2} \\
&= A(\theta)\cos(kr_{av} - \omega t) \qquad (7.41)
\end{aligned}
$$

where

$$A(\theta) = A\frac{\sin N\phi/2}{\sin \phi/2}$$

and recalling that

$$A = A(0)\frac{\sin \beta}{\beta}$$

is the amplitude due to diffraction from a single slit. Hence, $A(\theta)$ can explicitly be written in the form

$$A(\theta) = A(0)\frac{\sin[\frac{1}{2}kb\sin\theta]}{[\frac{1}{2}kb\sin\theta]}\frac{\sin[\frac{1}{2}Nkd\sin\theta]}{[\sin\frac{1}{2}kd\sin\theta]} \qquad (7.42)$$

The net intensity distribution observed on the screen is the square of the amplitude, given by

$$\begin{aligned} I(\theta) &= A^2(0)\frac{\sin^2[\frac{1}{2}kb\sin\theta]}{[\frac{1}{2}kb\sin\theta]^2}\frac{\sin^2[\frac{1}{2}Nkd\sin\theta]}{[\sin^2\frac{1}{2}kd\sin\theta]} \\ &= I(0)\frac{\sin^2\beta}{\beta^2}\frac{\sin^2 N\gamma}{\sin^2\gamma} \qquad (7.43) \end{aligned}$$

$$\text{where } I(0) = A^2(0) \qquad (7.44)$$

and one can identify

$$\frac{\sin^2\beta}{\beta^2} \qquad (7.45)$$

as the *diffraction term*, with $\beta = \frac{1}{2}kb\sin\theta$, and

$$\frac{\sin^2 N\gamma}{\sin^2\gamma} \qquad (7.46)$$

as the *interference term*, with $\gamma = \frac{1}{2}kd\sin\theta$.

A graph of the diffraction term is illustrated in Figure 7.12, while that of the interference term is illustrated in Figure 7.13 for $N = 6$. The products of these two graphs gives the diffracted intensity through the diffraction grating, and this is illustrated in Figure 7.14.

As in the case of the double slit diffraction, from Figures 7.12 to 7.14, we note that the diffraction pattern from the N-slits grating acts as an envelope of the interference pattern produces by the grating. Therefore, the locations of the maxima and minima in the resultant diffraction pattern are primarily determined by the interference term (equation 7.46).

The interference term is maximum, equal to one for $\gamma = m\pi$ where $m = 0, 1, 2, 3, ...$ is an integer. From this one obtaines the condition of mth order maximum as follows:

$$\begin{aligned} \frac{1}{2}kd\sin\theta &= m\pi = \frac{\pi}{\lambda}d\sin\theta \\ or \quad d\sin\theta &= m\lambda \end{aligned}$$

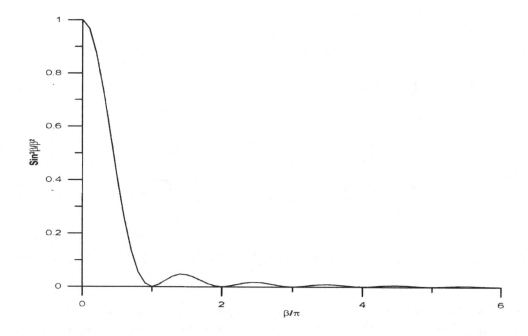

Figure 7.12: A graph of the diffraction term of a diffraction grating with $N = 6$.

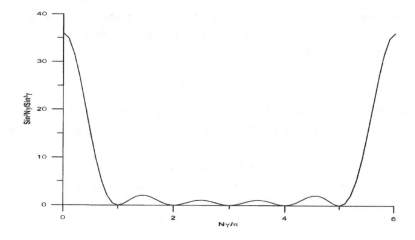

Figure 7.13: A graph of the interference term of a diffraction grating with $N = 6$.

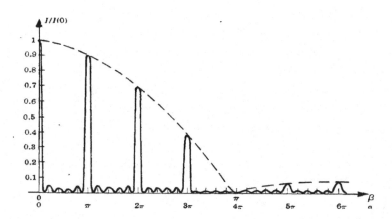

Figure 7.14: A graph of the intensity distribution of light diffracted by a diffraction grating with $N = 6$.

The zeroth order maximum is observed at $\theta = 0$, which is an obvious conclusion. d the spacing between the consequetive slits, also known as the *grating constant*, for a grating of length L with N number of slits is given by:

$$d = \frac{L}{N-1} \approx \frac{L}{N}, \quad for \quad large \quad N$$

The minimum of intensity are obtained when the diffraction term is zero, or when $sin\,(N\gamma) = 0$. This condition is satisfied for $N\gamma = n\pi$, where n is an integer excluding $n = 0, N, 2N, 3N, ...$ Explain why? The condition for minimum, thus reduces to:

$$N\,d\,sin\,\theta = n\,\lambda, \quad where \quad n \neq 0, N, 2N, 3N, ...$$

Lastly, the effect of the diffraction term is such that at angular positions where the minima of diffraction pattern coincide with the maxima of interference, one encounters *missing orders* of the spectrum.

Diffraction gratings have important applications in physics, for example in determination of wavelengths. In this regard, the concept of *dispersive power, D* is introduced, defined as

$$D = \frac{d\theta}{d\lambda} \tag{7.47}$$

For example, for $dsin\theta = m\lambda$, we obtain the dispersive power

$$D = \frac{d\theta}{d\lambda} = \frac{m}{d\cos\theta}$$

This equation has several factors of physical interest.

First, for a given small wavelength difference $d\lambda$, the angular separation $d\theta$ is *directly proportional* to the order m. Thus, the second order spectrum will be twice as wide as the first order spectrum.

Second, for a given small wavelength difference $d\lambda$, the angular separation $d\theta$ is *inversely proportional* to the slit separation d. Thus, the smaller the slit separation the wider the spectrum will be.

The effect of $\cos$ factor is that dispersion will be smallest when $\theta \approx 0$, that is close to the normal.

7.7   Fresnel diffraction by straight edges

Consider Fresnel diffraction by a system bound by straight edges, such as rectangular holes, slits, wires etc. The field at a point P is found by integrating all the differential contributions at the aperture.

From the geometry in Figure 7.15,

$$
\begin{aligned}
\rho^2 &= \rho_0^2 + y^2 + z^2 \\
r^2 &= r_0^2 + y^2 + z^2 \\
\rho &\approx \rho_0 + \frac{y^2 + z^2}{2\rho_0} \\
r &\approx r_0 + \frac{y^2 + z^2}{2r_0}
\end{aligned}
$$

and hence

$$
\rho + r \approx \rho_0 + r_0 + (y^2 + z^2)\frac{\rho_0 + r_0}{2\rho_0 r_0}
$$

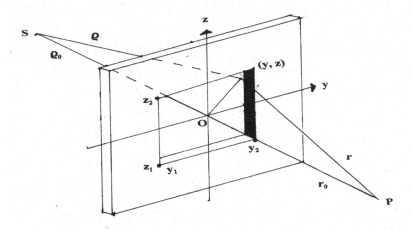

Figure 7.15: An illustration of Fresnel diffraction by straight edges.

The field at P is given by

$$E_P = \frac{E_0}{2(\rho_0 + r_0)} e^{i[k(\rho_0+r_0)-\omega t]} \int_{u_1}^{u_2} e^{i\pi u^2/2} du \int_{v_1}^{v_2} e^{i\pi v^2/2} dv \tag{7.48}$$

$$E_P = \frac{E_u}{2} \int_{u_1}^{u_2} e^{i\pi u^2/2} du \int_{v_1}^{v_2} e^{i\pi v^2/2} dv \tag{7.49}$$

where

$$\frac{E_u}{2} = \frac{E_0}{2(\rho_0 + r_0)} e^{i[k(\rho_0+r_0)-\omega t]} \tag{7.50}$$

$$u = y \left[\frac{2(\rho_0 + r_0)}{\lambda \rho_0 r_0} \right]^{1/2} \tag{7.51}$$

$$v = z \left[\frac{2(\rho_0 + r_0)}{\lambda \rho_0 r_0} \right]^{1/2} \tag{7.52}$$

The integral in equation (7.49) can be evaluated in terms of Fresnel integrals.

$$\int_{u_1}^{u_2} e^{i\pi u^2/2} du = \int_{u_1}^{u_2} \cos\left(\frac{\pi u^2}{2}\right) du + i \int_{u_1}^{u_2} \sin\left(\frac{\pi u^2}{2}\right) du \tag{7.53}$$

$$= C(u) + iS(u) \tag{7.54}$$

where

$$C(u) = \int_{u_1}^{u_2} \cos\left(\frac{\pi u^2}{2}\right) du \tag{7.55}$$

$$S(u) = \int_{u_1}^{u_2} \sin\left(\frac{\pi u^2}{2}\right) du \tag{7.56}$$

and similar expressions for the v integral. The field E_P can therefore be written in the form

$$E_P = \frac{E_u}{2} [C(u) + iS(u)]_{u_1}^{u_2} [C(v) + iS(v)]_{v_1}^{v_2} \tag{7.57}$$

Introducing $B(\omega)$ defined by

$$B(\omega) = C(\omega) + iS(\omega) \tag{7.58}$$

a curve of $S(\omega)$ against $C(\omega)$ is plotted on the complex plane for all values of ω, and illustrated in Figure 7.16. This curve is known as the *Cornu spiral*.

The field E_P for a rectangular aperture of straight edges is given by

$$E_P = \frac{E_u}{2} [B(u_2) - B(u_1)][B(v_2) - B(v_1)] \tag{7.59}$$

The diffracted intensity is the square of the electric field, given by

$$I_P = \frac{I_0}{2}|B_{12}(v)|^2$$

$$= \frac{I_0}{2}\left\{[C(v_2) - C(v_1)]^2 + [S(v_2) - S(v_1)]^2\right\} \qquad (7.60)$$

A particular case of interest is the Fresnel diffraction by semi-infinite straight edge, whose diffracted intensity distribution is given by

$$I_P = \frac{I_0}{2}\left\{\left[\frac{1}{2} - C(v_1)\right]^2 + \left[\frac{1}{2} - S(v_1)\right]^2\right\} \qquad (7.61)$$

and is illustrated in Figure 7.17.

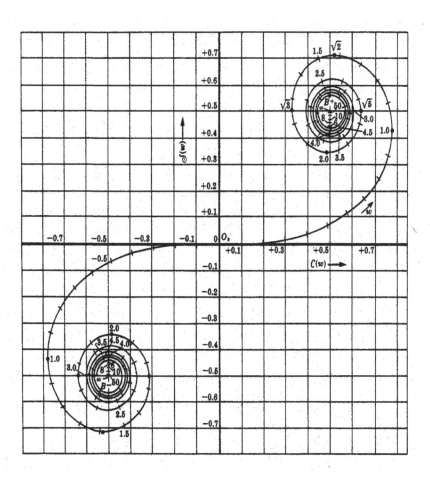

Figure 7.16: Cornu Spiral.

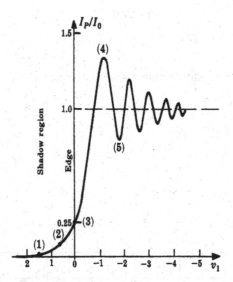

Figure 7.17: The intensity distribution in the diffraction pattern due to a straight edge. Points 1 and 2 are below the edge, 3 is at the edge, 4 and 5 are above the edge.

7.8 Diffraction by crystals

7.8.1 X-ray diffraction

X-rays can be diffracted by a crystal because the wavelength of the X-rays is of the same order of the magnitude as the interatomic spacing of the crystal. A schematic diagram of incident x-rays diffracted by crystal planes is illustrated in Figure 7.18. The fact that X-rays can be diffracted is a clear demonstration of their wave-like properties.

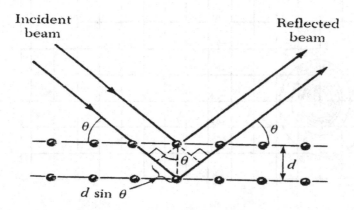

Figure 7.18: An illustration of X-ray diffraction from two parallel planes in a crystal. Note that the path difference between the two x-ray beams is $2d\sin\theta$.

The condition for the maxima of diffraction is given by Bragg's law, which states that the path difference between X-rays from two neighbouring planes must be an integer multiple of the wavelength.

$$2d\sin\theta = n\lambda \tag{7.62}$$

where
$n = 0, 1, 2, \cdots$
d is the distance between neighbouring planes
θ is the diffraction angle
λ is the wavelength
The distance d is calculated for several angles of diffraction, θ using Bragg's law. The lattice spacing a is calculated for several Miller indices (hkl) , for example, for crystals with a cubic structure, one obtains.

$$a = \frac{d}{\sqrt{h^2 + k^2 + l^2}} \tag{7.63}$$

From such observations, lattice structure and lattice parameters can be obtained. There are three techniques for X-ray diffraction (XRD), namely Laue method, Rotating Crystal method and Powder method that are widely employed to study crystals. Modern XRD machines are computer controlled, and have made XRD studies highly automated.

7.8.2 Theory of X-ray diffraction by crystals

Crystals are an ordered array of atoms of a solid. A crystal can be regarded as a 3-D diffraction grating, an analysis of which requires a rather complex mathematical treatment. Here, we present a simplified insight into the theory of x-ray diffraction by crystals. Crystals have what are known as a *direct lattice* and a *reciprocal lattice*, defined as below. X-rays map the reciprocal lattice.

If $\vec{a}_1, \vec{a}_2, \vec{a}_3$ are primitive vectors in the direct lattice, and $\vec{b}_1, \vec{b}_2, \vec{b}_3$ are primitive vectors in the reciprocal lattice, then

$$\vec{r}_j = x_j\vec{a}_1 + y_j\vec{a}_2 + z_j\vec{a}_3$$

is the direct lattice vector, with x_j, y_j and z_j being coordinates of atoms, and

$$\vec{G} = h\vec{b}_1 + k\vec{b}_2 + l\vec{b}_3$$

is the reciprocal lattice vector, with (hkl) being the Miller indices.

The *X-ray Scattering Amplitude*, $A(\vec{G})$ is given by

$$A(\vec{G}) = N\int_{\text{cell}} n(r)e^{-i\vec{G}.\vec{r}}dV \tag{7.64}$$

$$= NS_G \tag{7.65}$$

where S_G is the *Structure Factor*, N is the number of unit cells in the crystal, and $n(r)$ is the concentration of atoms within an individual unit cell.

$$
\begin{aligned}
S_G &= \int_{\text{cell}} n(\vec{r}) e^{-i\vec{G}.\vec{r}} dV \\
&= \sum_j \int n_j(\vec{r} - \vec{r}_j) e^{-i\vec{G}.\vec{r}} dV \\
&= \sum_j e^{-i\vec{G}.\vec{r}_j} \int n_j(\vec{r} - \vec{r}_j) e^{-i\vec{G}.(\vec{r}-\vec{r}_j)} dV \\
&= \sum_j f_j e^{-i\vec{G}.\vec{r}}
\end{aligned}
\tag{7.66}
$$

where

$$
f_j = \int n_j(\vec{r} - \vec{r}_j) e^{-i\vec{G}.(\vec{r}-\vec{r}_j)} dV
\tag{7.67}
$$

is the *Atomic Form Factor*.

$$
S_G \rightarrow F_{hkl}
$$

where F_{hkl} is the *Geometrical Structure Factor*.

$$
F_{hkl} = \sum_j f_j e^{-i2\pi(hx_j + ky_j + lz_j)}
\tag{7.68}
$$

The diffracted intensity by a crystal is proportional to the square of the X-ray scattering amplitude given by equation (7.64), or equivalently to the square of the Geometrical Structure Factor given by equation (7.68). A typical diffraction pattern by crystals is given in Figure 7.19 (After Nkoma and Ekosse).

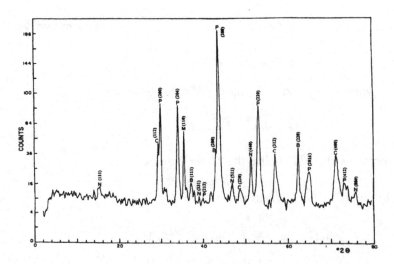

Figure 7.19: The X-ray Diffraction pattern of a sample of Cu-Ni orebody from Selebi-Phikwe, Botswana, showing peaks for pyrrhotite (P), pentlandite (N), chalcopyrite (C), magnetite (M) and bunsenite (B) (After Nkoma and Ekosse)

7.8.3 Electron diffraction

Electrons can be diffracted by a crystal because the de Broglie wavelength of the electrons is of the order of the magnitude of the interatomic spacing of the crystal. A pioneering study of electron diffraction was done by Davisson and Germer in 1927. The condition for electron diffraction is given by Bragg's law for *normal incidence*, as:

$$d \sin \theta = n\lambda \tag{7.69}$$

where λ is the de Broglie wavelength of the electron, given as $\lambda = h/p$, with p being the momentum of the electron, and θ is the diffrcation angle. If the electron is accelerated through a potential difference of V volts, then

$$\frac{p^2}{2m} = eV \tag{7.70}$$

where m is the mass of the electron, and hence

$$\lambda = \frac{h}{\sqrt{2meV}} = \frac{12}{(eV)^{\frac{1}{2}}}\mathring{A} \tag{7.71}$$

The fact that electrons can be diffracted is a demonstration of their wave-like properties. A typical electron diffraction pattern due to an aluminium film is illustrated in Figure 7.20, obtained using a Transmission Electron Microscope (TEM) (Courtesy: S H Coetzee, Electron Microscope Unit, Department of Physics, University of Botswana).

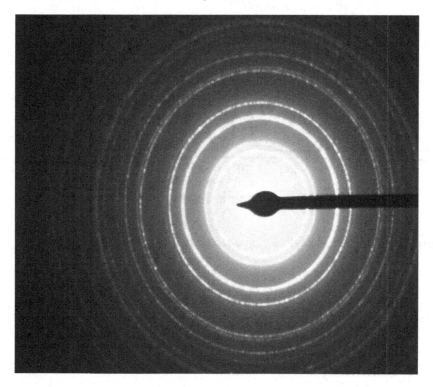

Figure 7.20: An electron diffraction pattern due to an aluminium film (Courtesy: S H Coetzee, Electron Microscope Unit, Department of Physics, University of Botswana).

7.8.4   Neutron diffraction

Neutrons can be diffracted by a crystal because the de Broglie wavelength of the neutrons is of the order of the magnitude of the interatomic spacing of the crystal. The condition for diffraction is given by Bragg's law for oblique incidence, as:

$$2d \sin \theta = n\lambda_n \tag{7.72}$$

where λ_n is the de Broglie's wavelength of the neutron, given as

$$\lambda_n = \frac{h}{p} = \frac{h}{\sqrt{2m_n E}} = \frac{0.28}{[E(eV)]^{\frac{1}{2}}}\mathring{A} \tag{7.73}$$

where p is the momentum of the neutron of mass m_n and energy E. The fact that neutrons can be diffracted is a demonstration of their wave-like properties. A typical neutron diffraction pattern due to diamond is illustrated in Figure 7.21. The different peaks are from different crystal planes with Miller indices (hkl).

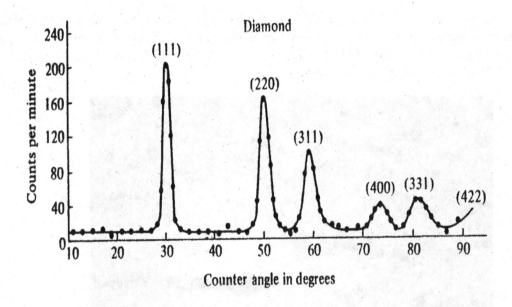

Figure 7.21: A neutron diffraction pattern due to powdered diamond.

The three radiations: X-rays, neutrons and electrons discussed in this chapter provide three distinctive windows to understanding crystal structure. X-rays are not charged and have no magnetic moment, neutrons are not charged but have a magnetic moment, electrons are negatively charged. Thus, these projectiles provide complementary information about crystals.

7.9   Chapter 7 Summary

Diffraction of light illustrates the bending of light from its rectilinear path around sharp edges, which can be explained in terms of the Huygens secondary wavelets. As a result the shadows/ images of objects do not have sharp edges as the rectilinear propagation of light shall produce, rather result in dark and bright fringe pattern, known as the diffraction pattern that extends beyond the sharp geometrical edges of the shadow/image. The effect can be observed only under controlled laboratory conditions using monochromatic source of light, and its study is divided in to two categories: Fraunhofer and Fresnel diffraction.

- Fraunhofer Diffraction: Both, the source of light and the observation screen are effectively located at infinite distance from the obstacle. Theory and characteristics of diffraction patterns from the following obstacles are presented.
  ○ Single slit: The central maxima, the geometrical image of the slit, is bounded by dark and bright fringe pattern. The dark fringes are located at angular positions, $\beta = \pm n\pi, (n = 1, 2, 3, \cdots)$. The maxima of diminishing intensity do not lie midway between the minima; their angular positions are given by: $\beta = \tan \beta$.
  ○ Rectangular aperture: The geometrical bright central image of the aperture is bounded by diffraction pattern of dark and bright fringes on all four sides. The conditions for the dark and bright fringes along the $x-$ and $y-$ dimensions of the aperture are same as for the single slit.
  ○ Circular aperture: Again, as expected the central bright image of the aperture is bounded by concentric pattern of dark and bright fringes extending in the geometrical dark region of the image. Resulting from the circular symmetry, the angular locations of the fringes are given by first order Bessel function.
  ○ Youngs Double slits: In the interference study, the diffraction by the slits was ignored. In reality the diffraction effect from a single slit acts as the envelope of the interference from both the slits. Thus the diffraction determines the intensity of the fringes which also results in missing fringes while interference determines the spacing of the fringes.
  ○ $N(> 2)$ slits, also known as the diffraction grating, is used in a number of precise optical studies and measurements. once again the resulting pattern is a product of the diffraction and interference effects. The fringe pattern has a number of order of spectra of diminishing intensity. The interference and diffraction effects are determined by similar mathematical terms as one encounters in the double slit fringe pattern.

- Fresnel Diffraction: Either the the source of light or the observation screen or both are located at finite distance from the obstacle. only one case of straight edge is presented. The intensity distribution in the diffraction pattern are fringes of sharply diminishing intensity, given mathematically by the Cornu Spiral.

- The chapter is concluded with the diffraction of x-rays, electron beam and the neutron beams. This is the direct evidence of the wave nature of the elementary particles such as electrons and neutrons, and are used in the study of crystal structure. The periodic atomic structure acts as a 3-D diffraction grating and the deBroglie wavelength of particles is of the same order as the atomic spacing. The x-rays, electrons waves, and neutron waves have differing physical properties, and are used to explore other properties of crystals as well.

7.10   Exercises

7.1. Fraunhofer diffraction is observed using a spectrometer illuminated with a sodium lamp (wavelength, $\lambda = 589$ nm) and with a slit of variable aperture width d inserted between the collimator and the telescope. The collimator lens has a focal length of 10 cm. If the collimator slit is replaced by double narrow slits separated by 1 mm, calculate the minimum width for which the double slits can be seen just separate when observed through the telescope.

7.2. A telescope has a diameter of 200 in. What is the limiting angle of resolution at a wavelength of 600 nm?

7.3. A collimated monochromatic beam (wavelength, $\lambda = 600$ nm) is incident normally on a converging lens of 1.2 cm diameter and focal length 50 cm. Calculate the diameter of the central disk on the focal plane.

7.4. A laser beam can be so well collimated that it spreads out only as a result of diffraction. Suppose such a laser beam of wavelength, $\lambda = 632.8$ nm has a diameter of 2 mm, what will be the beam diameter at a distance of 1 km from the laser?

7.5. By making reasonable approximations, show that the intensity of the mth maxima of the diffraction pattern through a single slit is given by

$$I_m = I(0) \left[\frac{1}{(m + \frac{1}{2}\pi)} \right]^2$$

and hence calculate the ratio of the intensity of the second maxima to the central maximum.

7.6. A 3 mm diameter hole in an opaque screen is illuminated by plane waves of wavelength 550 nm. Compute the locations of the first three maxima.

Exercises 7.7 to 7.10 require a knowledge of crystal structure.

7.7. (a) Show that for a bcc (body centred cubic) lattice, the geometrical structure factor is given by

$$F_{hkl} = f \left\{ 1 + e^{-i\pi(h+k+l)} \right\}$$

where all symbols are in the usual notation, and you are given that the bcc lattice has identical atoms at 000 and $\frac{1}{2}\frac{1}{2}\frac{1}{2}$.

(b) Hence, show that for a bcc F_{hkl} is equal to 0 or $2f$.

7.8. (a) Show that for a fcc (face centred cubic) lattice, the geometrical structure factor is given by

$$F_{hkl} = f \left\{ 1 + e^{-i\pi(k+l)} + e^{-i\pi(h+l)} + e^{-i\pi(h+k)} \right\}$$

where all symbols are in the usual notation, and you are given that fcc lattice has identical atoms at $000; 0\frac{1}{2}\frac{1}{2}; \frac{1}{2}0\frac{1}{2}; \frac{1}{2}\frac{1}{2}0$.

(b) Hence, show that for a fcc F_{hkl} is equal to 0 or $4f$.

7.9. Calculate the Bragg angle at which *electrons* accelerated from rest through a p.d of 80 V will be diffracted from the (111) planes of a f.c.c. crystal of lattice spacing 3.50A.

7.10. Calculate the effective temperature of thermal *neutrons* from a reactor if they are diffracted by the (111) planes of a Nickel crystal at an angle of $28°30'$ given that Ni is f.c.c. and has a lattice spacing of $3.52Å$.

7.11. Show that if the energy of electrons is given in eV, then the wavelength in $Å$ is given by

$$\lambda = \frac{12}{(eV)^{\frac{1}{2}}}Å$$

7.12. Show that if the energy of neutrons is given in eV, then the wavelength in $Å$ is given by

$$\lambda = \frac{0.28}{[E(eV)]^{\frac{1}{2}}}Å$$

7.13. Show that if the energy of X-rays is given in keV, then the wavelength in $Å$ is given by:

$$\lambda(Å) = \frac{12.4}{E(keV)} \tag{7.74}$$

7.14. Calculate the location and the order of the first three missing maxima on either side of the central maxima in the diffraction pattern of Young's double slit in terms of the slitd of width b and separation d.

7.15. A plane diffraction grating has a grating constant of $1.79 X 10^{-4}$ cm. Calculate the wavelength of monochromatic light for which the first and second order diffraction maxima are observed at $18°40'$ and $39°48'$ respectively.

7.16. Diffraction of light can be observed only when the size of the obstacle/ aperture is very small comparable to the wavelength of light. Discuss this statement considering the example of diffraction from a single slit.

7.17. Plane wave of $\lambda = 5 X 10^{-5}$ cm is incident normally on a slit of width 0.2 mm. Calculate the width of the central maxima observed on a screen placed 2.0 m away from the slit.

7.18. A plane diffraction grating has 5000 lines per cm. How many orders of diffraction spectra it shall produce for light of $\lambda = 5 X 10^{-5}cm$ at normal incidence. Calculate the wavelengths in the range 3500 to 8000 $Å$ which overlap with the third order spectra of the 5000 $Å$ wavelength.

7.19. With light of 5000 $\overset{\circ}{A}$, the angular width of the central maxima in single slit diffraction pattern is 40°. Calculate the width of the slit. What are the angular positions of the second and the third minima?

7.20. Using a grating of 5500 lines per centimeter, calculate the longest wavelength that will produce upto the third order spectrum.

7.21. A 2 cm long grating has 10, 000 lines. Calculate the angular separation between the two sodium lines in the second and third order spectra.

7.22. Calculate the number of lines per cm of a grating that will produce an angular separation of half a degree between the two sodium lines in the first order spectrum. What shall be their separation in the second order spectrum?

Chapter 8

Polarization of Light

8.1 Introduction

Vibrations of a transverse wave are perpendicular to the direction of wave propagation. The most commonly known transverse waves are the vibrations of a stretched string. The vibrations of such waves can have any random orientation in the plane perpendicular to the direction of propagation of the wave (Figure 8.1 a). If one looked at the wave along the direction of propagation displacement due to vibrations shall be seen distributed around the direction of vision. Such a wave motion is called an *un-polarized* wave. However, if by some means we can restrict the vibrations of the wave in only one plane containing the direction of propagation of the wave, then the wave is called *plane polarized* wave (Figure 8.1 b), and the plane containing the vibrations is known as the *plane of polarization*.

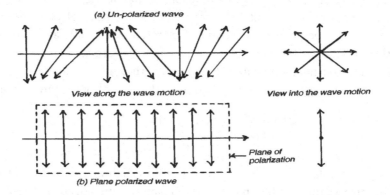

Figure 8.1: (a) Unpolarized, and (b) polarized wave motions as viewed along the direction of propagation and into the direction of propagation.

As a demonstration of polarization, consider waves on a stretched string, one end of which is fixed and the other end is oscillated sideways in random directions (Figure 8.2). The transverse displacement of the string shall be randomly oriented as in the portion *AB* of the string. This portion of the string

177

represents unpolarized waves propagating along the string. At *B* the string passes through a vertical slit S_1. The slit allows only those vibrations of the string that are in the plane of the slit, and the components of other vibrations parallel to it to pass through to the other side. Waves in portion *BC* of the string have displacement in the plane of slit S_1. These are plane polarized waves, with plane of polarization in the plane of the slit. Slit S_1 by which polarization is achieved is called the *polarizer*. At *C* the string passes through another, identical slit S_2. If slit S_2 is parallel to slit S_1 the plane polarized waves in section *BC* shall pass through it, and if S_2 is perpendicular to S_1 as shown in the figure waves will not be allowed to pass beyond *C*. At any other orientation component of the plane polarized wave parallel to S_2 is allowed to pass through. Thus slit S_2, known as the *analyzer*, acts as the detector of plane polarized waves, and with its help one can determine the plane of polarization of the wave. From this demonstration we also note that the polarizer and the analyzer are generally identical devices, and they derive their name from the application for which they are used.

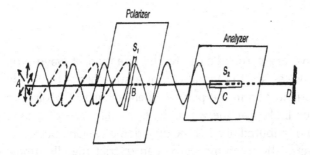

Figure 8.2: Polarization and analysis of waves on a stretched string.

When the analyzer is rotated about the polarized wave, the intensity of the wave passing through the analyzer varies between maximum to zero. For two positions of the analyzer when it is *parallel* to the polarizer which we shall denote as $\theta = 0$ and 180° positions the intensity of polarized wave passing through it is maximum. For two perpendicular positions, denoted by $\theta = 90^o$ and 270° positions the intensity is zero. In this orientation the polarizer and the analyzer are said to be *crossed*. If I_{max} and I_{min} are the maximum and minimum intensities passing through the analyzer for the parallel and crossed orientations respectively, we define percentage (degree of) polarization *P* as:

$$P = \frac{I_{max} - I_{min}}{I_{max} + I_{min}} \times 100\% \qquad (8.1)$$

From equation (8.1) if $I_{min} = 0$, beam is said to be pure (100 %) plane polarized, for any other value of $I_{min} \neq 0$ it is partly polarized or elliptically polarized which are discussed later. If there is no variation in intensity on rotation of the analyzer the beam is unpolarized or circularly polarized also discussed later.

In a longitudinal wave, such as the compressional waves propagating along the length of a spring or the sound waves, the displacement is along the direction of propagation of the wave. These vibrations can not be restricted by any means without restricting the wave itself. Thus polarization has no relevance to longitudinal waves, and polarization is a sure definite, and the only evidence of the transverse nature of waves. Light, that are electromagnetic waves, exhibit polarization and are, therefore, transverse in nature.

8.2 Polarization of light

Once again, we shall deal with light (electromagnetic waves) in terms of the electric field vector $\mathbf{E}$(x, t). In unpolarized light, the electric field vector has every possible orientation in the plane perpendicular to the direction of propagation. The unpolarized light, therefore, when viewed along and into the direction of propagation appears as shown in Figure 8.1(a), where the vibrations are oscillating $\mathbf{E}$(x, t). Consider two mutually perpendicular planes containing the direction of propagation, one being the plane of incidence coinciding with the plane of paper, and the other perpendicular to the plane of incidence. All the electric field vectors of unpolarized light can be resolved into components lying in these two planes (Figure 8.3a), and unpolarized light can be visualized as consisting of electric field vectors lying in the plane of incidence and perpendicular to the plane of incidence as shown in Figure 8.3(b). The dots along the direction of propagation represent the electric field vectors in the perpendicular plane. Electric field vector of plane polarized light lie only in one plane, which could either be the plane of incidence or the perpendicular plane. This gives two cases of plane polarized light: *(i)* polarized in the plane of incidence (Figure 8.3c) (also known as *p-waves*), and *(ii)* polarized perpendicular to plane of incidence (Figure 8.3d)(also known as *s-waves*).

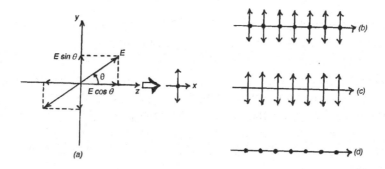

Figure 8.3: (a) Resolution of an electric field vector. (b) Unpolarized light. (c) Plane polarized light in the plane of incidence, and (d) Plane polarized light perpendicular to the plane of incidence.

8.2.1 Circular and elliptical polarized light

Recalling Lissajous figures, when two simple harmonic motions perpendicular to each other having same frequency and amplitude are superimposed, the resultant motion depends on the phase difference between the two motions, and the trajectory of the point on which the two motions are imposed is given by the Lissajous figure. If the phase difference between them is 0 or π the resultant motion is along a straight line, if the phase difference is $\pi/4$ or $3\pi/4$ the resultant motion is elliptical, and if the phase difference is $\pi/2$ the resultant motion is circular (Figure 8.4). If the amplitudes of the two motions are unequal, then with a phase difference of $\pi/2$ also the resultant motion is elliptical. In fact, a circle may be regarded as a special case of an ellipse whose major and minor axes are equal.

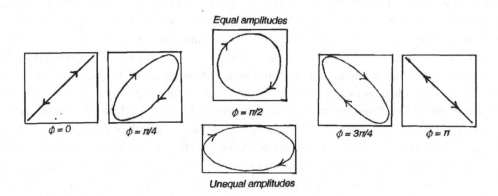

Figure 8.4: Lissajous figures from the superposition of two perpendicular harmonic motions of equal frequencies and amplitudes with a phase difference of ϕ.

Likewise, when two plane polarized light waves of equal amplitude and frequency with their polarization perpendicular to each other are superimposed, the resultant wave motion is plane polarized if the phase difference between them is 0 or π, elliptically polarized if the phase difference is $\pi/4$ or $3\pi/4$, and circularly polarized if the phase difference is $\pi/2$. Here we shall only consider two cases of superposition of two perpendicularly plane polarized light waves with a phase difference of $\pi/2$ having equal and unequal amplitudes. Let x be the direction of wave propagation, and y and z and the directions of the electric field vectors of two perpendicularly plane polarized waves with amplitudes E_{oy} and E_{oz} respectively. If the phase difference between them is $\pi/2$, then the corresponding field vectors at time t at location x are:

$$E_y = E_{oy}sin(\omega t - kx) \tag{8.2}$$
$$E_z = E_{oz}sin\left(\omega t - kx + \frac{\pi}{2}\right) = E_{oz}cos(\omega t - kx)$$

from equation (8.2) we get:

$$\frac{E_y^2}{E_{oy}^2} + \frac{E_z^2}{E_{oz}^2} = 1 \tag{8.3}$$

equation (8.3) is the equation of the resultant wave motion for all values of x and t. This is the equation of an ellipse *i.e.*, the resultant wave motion is elliptically polarized for all values of time t and at all point along the direction of propagation. E_{oy} and E_{oz} are the two axes of the ellipse.

If the amplitudes of both the waves are equal, *i.e.*, $E_{oy} = E_{oz} = E_o$, the resultant wave motion becomes:

$$E_y^2 + E_z^2 = E_o^2 \tag{8.4}$$

This is the equation of a circle, *i.e.* the resultant wave motion in this case is circularly polarized.

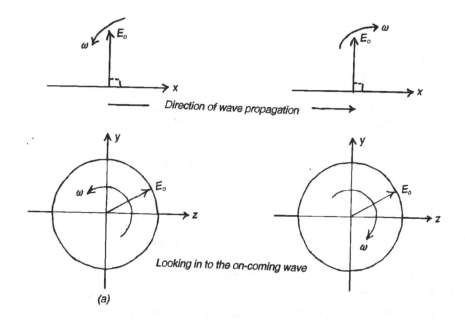

Figure 8.5: (a) Positive, and (b) negative hellicity of circularly polarized light.

In circularly polarized light a constant electric field vector equal to the amplitude of individual waves (E_o) rotates in a circle around the direction of propagation with the same frequency as that of the wave. There are two possible directions of rotation: looking into the oncoming wave, *(i)* if the **E** vector rotates counter-clockwise, the wave is called *left circularly polarized*, and is said to have *positive helicity*. *(ii)* If the rotation of the **E** vector is clockwise, it is called *right circularly polarized*, and has *negative helicity* (Figure 8.5). Projections of the **E** vector along the *y* and *z* axes give the instantaneous electric field vectors of the two constituent perpendicularly plane polarized waves.

In elliptically polarized light also the electric field vector rotates in the same way as for the circularly polarized light, but its magnitude varies between the lengths of the semi-major and semi-minor axes of the ellipse.

8.3 Production of plane polarized light

Plane polarized light can be produced by any one of the following means.

- Reflection (refraction)
- Double refraction
- Selective absorption
- Polaroid sheets
- Scattering

8.3.1   Polarization by reflection (refraction)

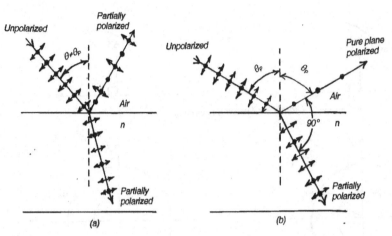

Figure 8.6: (a)Polarization by reflection, and (b) Brewster's angle.

When a beam of unpolarized light is reflected from a dielectric medium (such as glass or water), there is a preferential reflection of waves with electric field vector perpendicular to the plane of incidence (parallel to the reflecting surface), whereas the electric filed vector in the plane of incidence has lower reflectance. Thus the reflected beam is partially polarized with abundance of electric field vectors parallel to the reflecting surface (Figure 8.6 a). The degree of polarization depends on the angle of incidence. The polarization is zero, *i.e.*, the reflected beam is completely unpolarized at normal and grazing incidence ($i = 0^o$ and $\sim 90^o$), it is pure polarized at an angle of incidence $i = \theta_p$, known as the *Brewster's angle* (Figure 8.6 b), and at any other angle of incidence the polarization of the reflected beam is partial. At Brewster's angle of incidence the reflected and the refracted rays are perpendicular to each other. If r is the angle of refraction, and assuming the refractive index of air to be 1:

$$\theta_p + 90^o + r \;=\; 180^o, \quad or \quad r = (90 - \theta_p),$$

$$and, \quad n \;=\; \frac{sin\,\theta_p}{sin\,(90 - \theta_p)} = tan\,\theta_p \tag{8.5}$$

Equation (8.5) is known as the *Brewster's Law*. As refractive index depends on the wavelength of light, so does the Brewster's angle.

After one reflection, the intensity of the reflected beam is very weak, as most of the intensity of the incident wave is refracted into the medium. Therefore, even though the reflected beam at Brewster's angle is pure plane polarized, it does not provide enough intensity for practical application. In order to obtain higher intensity one uses a pile of identical glass plates (same refractive index). On each incidence (from both surfaces of each plate), more plane polarized waves are reflected as shown in Figure 8.7, which add up to give a plane polarized beam of larger intensity.

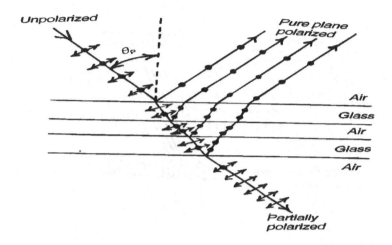

Figure 8.7: Producing a plane polarized beam of large intensity by reflection from a stack of glass plates.

Although there is no preferential refraction of waves with electric field vectors in any particular plane, because of the polarization of the reflected beam the refracted beam is also partly polarized with more electric field vectors in the plane of incidence than in the perpendicular plane. The polarization of the refracted beam remains partial even at Brewster's angle. But when a number of glass plates are used to obtain a plane polarized reflected beam of large intensity, the refracted beam also gets more and more polarized. In this way the transmitted beam from a stack of glass plates approaches 100% polarization with electric field vectors in the plane of incidence. If the intensities of waves with electric field vectors in the plane of incidence and perpendicular to it are $I_{\parallel}$ and $I_{\perp}$, the degree of polarization of the transmitted beam, from equation (8.1) is given by:

$$P = \frac{I_{\parallel} - I_{\perp}}{I_{\parallel} + I_{\perp}} \times 100\% \tag{8.6}$$

In terms of the number of plates m (*2m* reflecting surfaces) and their refractive index *n*, considering the contribution of rays that undergo multiple internal reflections in each plate, the polarization of the refreacted beam is given by:

$$P = \frac{m}{m + (2n^2/(1+n)^2)} \times 100\% \tag{8.7}$$

For common glass of refractive index $n = 1.5$, the refractive index term in the denominator in equation (8.7) is ≈ 0.7. Thus with one plate one would get about 58% polarization of the transmitted wave, and with $m = 10$ it approaches 93%. Figure 8.8 shows stack of plates mounted at Brewster's angle, used as a polarizer and analyzer by refraction of light.

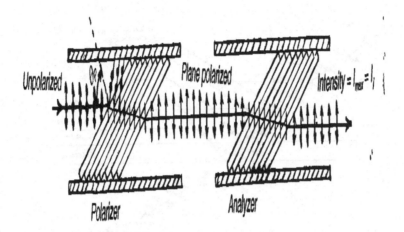

Figure 8.8: Stack of glass plates mounted at Brewster's angle used as a polarizer and analyzer.

8.3.2 Polarization by double refraction

Double refraction

When a crystal of calcite (hydrated $CaCO_3$ also known as Iceland spar) is placed on a flat object such as a printed sheet of paper, one sees two images of the print on the page. An investigation of these images reveals the following refraction properties of the crystal:

- When the crystal is rotated about a vertical axis, one of the images remains fixed, while the other image rotates along a circle about the fixed image. The ray which produces the fixed image is known as the *ordinary ray* or the *O-ray* and the ray that produces the rotating image is known as the *extraordinary ray* or the *E-ray*

- If the two images are examined through a Polaroid film (types of polarizers/ analyzers discussed later in section 8.3.4), at certain orientation of the film one of the image disappears and only one image is seen. On rotating the Polaroid sheet by 90° about the line of vision, the image that was seen earlier disappears, and the other image appears. Thus on rotating the Polaroid film, each of the two images appear and disappear alternately for every 90° rotation of the film. From this observation one concludes that each of the two images are formed from pure plane polarized beams, and the polarization of the two beams is perpendicular to each other.

Crystals (materials) that have these refraction properties are known as *double refracting* or *bire-fringent* crystals. Other examples of similar double refracting crystals are quartz, sodium nitrate ($NaNO_3$), ice and tourmaline. Although these materials are isotropic in their chemical composition, the very fact that an incident beam on refraction through them splits into two beams of different characteristics implies that they are optically anisotropic. From further investigation of the optical properties of these crystals it is found that for O-ray the crystal is optically isotropic, the speed of propagation of the O-ray, v_o, is the same along every direction in the crystal. The refractive index corresponding to the O-ray, $n_o = (c/v_o)$, has a constant value, and it can be expressed in terms of the angles of incidence and refraction as $n_o = (sin\,i/sin\,r_o)$, *i.e.*, the O-ray obeys the Snell's law. In contrast to this the crystal is optically anisotropic for the E-ray, speed of the E-ray v'_e is different along different directions. The refractive index of the crystal for E-ray, $n'_e = c/v'_e$, varies with the direction of the propagation of the wave. The E-ray does not obey the Snell's law *i.e.* refractive index can not be expressed in terms of the angles of incidence and refraction ($n'_e \neq (sin\,i/sin\,r'_e)$ in general). It is from this feature the two rays derive their names as O- and E- rays respectively.

Along one of the direction in the crystal, the speeds of both O- and E- rays are the same and so are their refractive indices. This direction in the crystal is known as the *optic axis* of the crystal, (and all directions parallel to an optic axis are also optic axes.) Along other directions, depending on the material, the speed of the E-ray is either smaller or larger than that of the O-ray, and along perpendicular to the optic axis the speed of E-ray reaches a minimum or a maximum value. The principal refractive index for the E-ray is given in terms of its speed perpendicular to the optic axis, *i.e.* $n_e = c/v_e$. In crystals with $v_e > v_o$, the refractive index $n_e < n_o$, and if $v_e < v_o$, the refractive index $n_e > n_o$. Crystals with $n_e < n_o$ are termed as *negative* crystals ($n_e - n_o$ is negative), and those with $n_e > n_o$ are called *positive crystals* ($n_e - n_o$ is positive). Table 8.1 gives the reflective indices for O-ray and E-ray for different materials for the sodium light. The first three materials in the table are negative, and the later two are positive birefringent. We also note that the difference in refractive indices $n_e - n_o$ is large for calcite and $NaNO_3$ compared to the other three materials. These materials which are both negative crystals, are more strongly anisotropic and show strong birefringence as compared to the others.

Table 8.1: Refractive indices of some birefringent crystals for the sodium light.

Material	n_o	n_e
Calcite	1.65836	1.48641
$NaNO_3$	1.5874	1.3361
Tourmaline	1.659	1.638
Quartz	1.5442	1.5534
Ice	1.3091	1.3104

The doubly refracting crystals are also classified as *uniaxial* and *biaxial* crystals. In the crystals given in Table 8.1, the O-ray obeys Snell's law while the E-ray does not, and there is one optic axis. They are known as the *uniaxial* crystals. In biaxial crystals none of the rays obeys Snell's, and there are two optic axes at an angle to each other which is characteristic of the material, and depends of wavelength. Uniaxial crystals can be visualized as a special case of biaxial crystals in which the angle between the two optic axes is zero. In this chapter we shall restrict our study to uniaxial, birefringent materials only.

Huygen's treatment of double refraction

Consider a point source of light embedded inside a double refracting crystal. The Huygen's wavefront for the O-ray is spherical, and that for the E-ray is ellipsoid. Along the optic axis the two wave fronts touch each other (why?). In positive crystals the major axis of the ellipsoidal wavefront of E-ray is along the optic axis and the ellipsoidal wavefront of the E-ray is contained inside the spherical wavefront of the O-ray (Figure 8.9 (a)). In negative crystals the spherical wavefront of the O-ray is contained inside the ellipsoidal wavefront of the E-ray (Figure 8.9 (b)). Explain why.

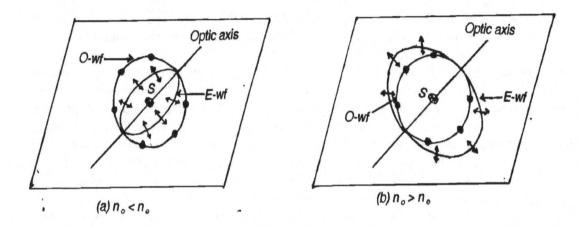

Figure 8.9: Huygen's wavefronts for the O- and E- rays from a point source embedded in a bire-fringent crystal with (a) $n_o < n_e$, (positive crystal), and (b) $n_o > n_e$ (negative crystal).

We now consider an unpolarized parallel beam of light incident at an angle i on a calcite crystal. It is a negative crystal ($n_o > n_e$), and the spherical O-wavefront is contained inside the ellipsoidal E-wavefront. Refraction of the O- and E- rays in the crystal depends on the orientation of the optic axis. We deal here with two cases as shown in Figures 8.10 (a) and (b). In Figure 8.10 (a) the optic axis is parallel to the crystal surface and the plane of incidence, and in Figure 8.10 (b)the optic axis is perpendicular to the refracting surface.

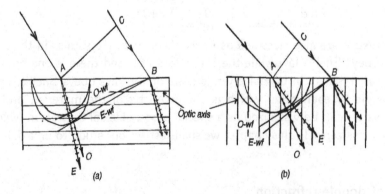

Figure 8.10: Refraction in a calcite crystal from Huygen's wave theory for optic axis (a) refracting parallel to the surface of the crystal and parallel to the plane of incidence and (b)perpendicular to

the refracting surface of the crystal.

The Huygen's wave fronts for the O- and E- rays, and corresponding refracted rays for both the cases are plotted following the procedure in section 2.4.1. From these figures one notes an interesting feature of double refracting crystals, stated earlier as a property of birefringent uniaxial crystals. In calcite crystal the refractive index n'_e for the E-ray in any direction other than the optic axis is less than the refractive index for the O-ray. Yet the angle of refraction r_e for the E-ray in Figure 8.10 (a) is smaller than that for the O-ray, which from Snell's law shall wrongly yield $n'_e > n_o$. This is so because the E-ray does not always lie in the plane of incidence, and its angle of refraction is not generally related to the angle of incidence through Snell's law. This emphasizes the earlier statement that the E-ray does not obey Snell's law, and the refractive index for E-ray in any direction can only be obtained from $n'_e = c/v'_e$. As for the O-ray, it always lies in the plane of incidence, and the wavefront being spherical, the angle of refraction in both cases of Figure 8.10 is the same. The refractive index for O-ray can be expressed in terms of the Snell's law as $n_o = sin\ i/sin\ r_o$. In this context, it is appropriate to conclude that fundamentally the refractive index of a material is defined in terms of the speed of light in the medium, and Snell's law is a special case for optically isotropic materials for which the refracted ray always lies in the plane of incidence.

If the light is incident normally, the refracted O- and E- rays in both cases travel in the same direction along the normal. For the orientation of the optic axis shown in Figure 8.10 (a) the E-ray travels faster than the O-ray, and a phase difference is introduced between them as they pass through the crystal. This property is applied in section 8.4 to construct quarter and half wave plates to produce elliptical and circular polarized lights. For the orientation of the optic axis shown in Figure 8.10 (b) both O- and E- waves travel with the same speed and no phase difference is introduced between them. Students are assigned to draw refraction diagrams similar to Figures 8.10 (a, b) for normal incidence.

Calcite crystal (Hydrated CaCO$_3$

In this chapter we restrict ourselves to the calcite crystal only; discuss the crystal parameters, direction of the optic axes, planes of polarization of O- and E- rays, and its application for the polarization of light. Calcite crystal is rhombohedral crystal, each face of which is a similar parallelogram with angles $78^o5'(\sim 78^o)$ and $101^o55'(\sim 102^o)$ as shown in Figure 8.11 (a).

All three crystal faces meeting at two opposite corners *A* and *C* along the body diagonal have obtuse angle (102^0). These corners of the crystal are termed as the *blunt corners*. The optic axis passes through the blunt corner such that it makes equal angles with all the faces meeting at the corner. A plane through the crystal containing the optic axis and normal to the opposite crystal faces is known as the *principal section*. Corresponding to the three pairs of crystal faces there are three distinct principal sections, and any plane parallel to a principal section is also a principal section. A principal section cuts the crystal in a parallelogram with angles 71° and 109° as shown in Figure 8.11 (b). In a crystal of equal breadth and width, for example if sides *AF* and *AG* of the crystal in Figure 8.11(a) are equal, the principal section through the top and bottom faces passes through the short diagonals *AB* and *DC* of these faces. The optic axis in the principal plane makes an angle of 45° with the short diagonal *AB*. An unpolarized ray of light *I* incident on face *AB* of the crystal on refraction is separated into perpendicularly plane polarized O- and E-rays. The electric field vector of the E-ray lies in the principal plane, parallel to the short diagonal *AB*, and for the O- ray it is perpendicular to the principal plane. Since the two opposite faces of the crystal are parallel, the emergent E- and O-rays are parallel to the incident ray.

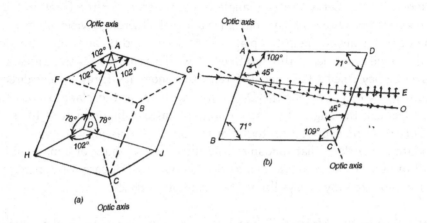

Figure 8.11: (a)A calcite crystal, and (b) its principal section.

Nicol Prism

In order to produce plane polarized light using calcite crystal, it is cut, polished, and rejoined in such a manner that one of the plane polarized ray is eliminated by total internal reflection, and only the other plane polarized ray emerges from the device. One such device is the *Nicol prism*. In fact the Nicol prism is not a prism as such, and it would be more appropriate to call it a Nicol rhombo, but the name prism is given perhaps because it is made of two slant prisms on rhombus base. A Nicol prism is made from a calcite crystal in the following manner:

- We begin with a calcite crystal of equal width and breadth, and nearly three times length *i.e.*, $AF = AG$ and $AD \approx 3 \times AF$ as shown in Figure 8.12 (a).

- Thin slant slices of the crystal are cut off from the two end faces of the crystal so that the angle of the principal plane is reduced from 71° to 68° as shown by the dashed lines in the figure. The cut faces are polished.

- the crystal is cut into two prisms by plane *A'KC'L*. The plane *A'KC'L* is normal to the short diagonals *A'B* and *C'D* of the cut end faces.

- The cut surfaces are polished and then cemented together by a transparent cement, *canada balsam*

- The four long sides of the cemented crystal are blackened to prevent stray radiation entering the prism.

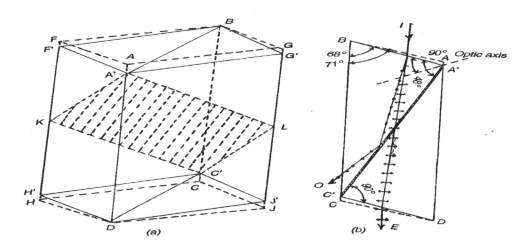

Figure 8.12: (a) Cuts of a calcite crystal to make a Nicol prism, and (b) its principal section and refraction through the Nicol.

The refractive index of canada balsam is 1.55 which is lower than the refractive index of calcite for the O-ray and larger than that for the E-ray. When an unpolarized ray of light is incident on the prism parallel to the long edge BC', it is split into O- and E-rays on refraction. The O-ray is incident at the first calcite-canada balsam interface at an angle greater than the critical angle, and is totally internally reflected. The E-ray is refracted through the canada balsam layer into the other section of the prism, and emerges from the opposite face of the Nicol, parallel to its long edge as a plane polarized light with electric field vector parallel to the short diagonal $A'B$. The principal section of the Nicol, and refraction through it are shown in Figure 8.12(b). The optic axis makes an angle of 48° with the short diagonal.

A number of other polarizing optical devices are made from doubly refracting crystals which include *Glan-Thompson prism* which has a large aperture of the order of 40°, *Foucault prism* that transmits ultraviolet light, and the *Rochon* and *Wollaston* prisms through which both perpendicularly polarized beams are transmitted, and are separated by an angle. Details of the construction of these devices is left as an independent study exercise for students.

Measuring the refractive index of birefringent crystals

As we noted in section 3.2.2, the refractive index of birefringent crystals, particularly for the E-ray can not be measured simply by measuring the angle of refraction because it does not obey the laws of refraction. Instead we determine the refractive index from the angle of minimum deviation using prisms made of these materials. The prisms for this purpose are made with two possible orientations of the optic axis as shown in Figs. 8.13 (a) and (b). In Figure 8.13 (a) the optic axis is perpendicular to the base of the prism, and in Figure 8.13 (b) the optic axis is parallel to the refracting edge of the prism. For both corresponding prisms and at minimum deviation for each ray, the corresponding ray through the prism travels parallel to the base of the prism which is perpendicular to the optic axis. Hence in this case, Snell's law can be applied to both the rays, and refractive indices determined from the corresponding angles of minimum deviation D_o and D_e as:

$$n_o = \frac{\sin\left(\frac{A+D_o}{2}\right)}{\sin\left(\frac{A}{2}\right)}$$

$$n_e = \frac{\sin\left(\frac{A+D_e}{2}\right)}{\sin\left(\frac{A}{2}\right)} \tag{8.8}$$

where A is the angle of the prisms.

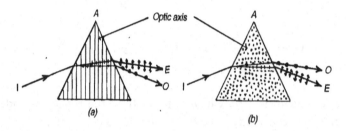

Figure 8.13: Prisms of a birefringent crystal with two orientations of the optic axis.

8.3.3 Polarization by selective absorption

Certain double refracting crystals, such as the mineral crystal *tourmaline*, selectively and completely absorb one of the plane polarized waves while the other plane polarized wave is transmitted without loss of intensity. Thus a beam of unpolarized light after passing through a thin plate of the crystal is 100% plane polarized. This property of crystals is known as *dichroism* and such crystals are known as *dichroic* crystals. Figure 8.14 shows three tourmaline crystals with a beam of unpolarized light incident from the left on the first crystal which acts as the polarizer.

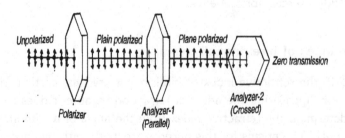

Figure 8.14: Dichroic property of tourmaline crystals used to produce a beam of plane polarized

light and to analyze it.

On passing through the first crystal, the O-wave is absorbed, and the E-wave is transmitted. The transmitted wave is plane polarized with electric field vector parallel to the long edge of the crystal. The second crystal used as analyzer-1 is parallel to the polarizer, and it allows the polarized light to pass through it. The third and the last crystal, used as analyzer-2 is rotated by 90°, *i.e.,* it is in crossed position relative to the polarizer, and no light is allowed to pass through it. Tourmaline crystals have a tint of colour, and therefore, they are not used in optical instruments as polarizer and analyzer.

8.3.4 Polarization using Polaroid sheets

Large area dichroic crystals and doubly refracting prisms are at times difficult to produce and are costly for commercial applications. Polaroid are the large area sheets of plastic materials produced commercially with dichroic property. The first Polaroids were manufactured in 1932 with thin sheets of nitrocellulose, heavily embedded with ultramicroscopic crystals of an organic compound known as quinine iodosulphate (also known as herapathite). The crystals absorb one of the polarized beam completely, while the other is transmitted with nearly no loss. The optic axis of herapathite crystals are aligned along one direction in the plane of the sheet, and the Polaroid sheet so produced provides plane polarized light with electric field vector perpendicular to the line of alignment of the crystals. Figure 8.15 shows Polaroid films used as polarizer and analyzer in parallel and crossed positions.

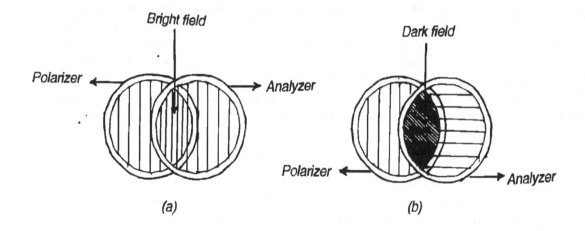

Figure 8.15: Polaroid polarizer and analyzer in (a) parallel, and (b) crossed positions.

Polaroid sheets used these days are manufactured by stretching plastic sheets containing long hydrocarbon chains, such as polyvinyl alcohol (PVA) so that the chains align along one direction. The sheet is then dipped in an iodine solution, and iodine gets attached to the long aligned chains of PVA. It is the long parallel strings of iodine atoms that impart dichroic property to the Polaroid sheets. The electric field vector parallel to the iodine chains are absorbed, and the perpendicular vibrations are transmitted. These Polaroids are known as the H-Polaroid. In yet another type of Polaroids known as the K-Polaroid, when a PVA sheet is heated in the presence of an active dehydrating catalyst such as hydrogen chloride, the film darkens slightly, and becomes strongly dichroic. K-Polaroids are very stable, as they do not have any dye material embedded in them, and are not bleached in sunlight. They are used for automobile headlights, and visors. Polaroids films are generally sandwiched and sealed between thin glass plates to protect the plastic sheet from scratches and damage. Polaroid sheets do not polarize all the wavelengths fully. Some extreme red and ultraviolet radiations pass through unpolarized.

8.3.5   Polarization by scattering

When light passes through a medium containing fine particles such as suspended dust particles in air, it undergoes scattering. The magnitude of scattering suffered by light is proportional to the fourth power of the inverse of wavelength, referred to as the fourth-power law. Thus light of shorter wavelength, towards the blue end of spectrum suffers more scattering, and long wavelength light towards the red end suffer much less scattering. From the fourth-power law and the average wavelengths of red and violet lights it is easy to deduce that for 10 waves of violet light scattered, only one red wave undergoes scattering.

Scattering in nature

It is the large scattering of blue component from the sun's white light by the earth's atmosphere which is responsible for the blue colour of sky, and for the red colour of the setting or rising sun on not so clear a day. The sky has no luminescence of its own. During the day it is full of blue scattered light which gives it the luminescence and the colour we see. If there was no scattering, the sky would have been pitch dark just like the sky at night. It would have been lit only where there were direct sun rays, like shining a flash light in pitch darkness. At sunrise and sunset, light from the sun passes through a larger length of earth's atmosphere, as a result of which much more blue component is removed from it by scattering. Hence, light from the rising and setting sun has larger red component which makes it appear red. Sun set or sunrise on a very clear day or after the rains does not appear that much red, because of the presence of less scattering dust particles in the atmosphere.

Polarization of scattered light

If the blue sky is viewed through a Polaroid film at different angles to the incoming sunlight, brightness of the sky is found to vary between a maximum to minimum for two perpendicular positions of the film because the scattered light is partly plane polarized. A careful investigation with Polaroid film further reveals that the maximum polarization of scattered light is at 90° to the direction of the incoming sunlight. Polarization by scattering occurs through the following mechanism.

When the electric field vector of incident light interacts with the micro dust particles in air, electrons in the micro particles are set in oscillation parallel to the direction of the incident electric field vector. A vibrating electron acts as an oscillating dipole, and emits electromagnetic radiation in every possible direction except along the direction along which it is vibrating. Thus the re-emitted (scattered) radiation from a vibrating electron lack radiation with the electric field vector perpendicular to its direction of vibration, and is therefore partially plane polarized.

Consider an unpolarized light traveling along the x- axis through a cloud of dust (Figure 8.16). The electric field vectors in the incident light have components along the y- and z- axes, which will set the electrons in the dust particles vibrating along y- and z- axes respectively. The electrons vibrating along the y- axis re emit radiation in every possible direction except along the y- axis, and such radiation has electric field vectors in every possible direction except in the plane perpendicular to the y- axis. In the same way, electrons set in vibration by the incident electric field vector along the z- axis do not produce scattered radiation with the electric field along the z- axis, and the scattered radiation shall lack radiation with electric field vectors perpendicular to z- axis. A collective effect of these processes is the scattered light which is maximum polarized at right angles to the direction of the incident light.

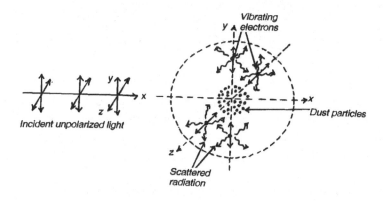

Figure 8.16: Scattering of light by dust particles.

8.4 Production of circular and elliptically polarized light

As discussed in section 8.2.1, circular and eliptically polarized waves are obtained from the superposition of two perpendicularly polarized waves of same frequency with equal or unequal amplitudes respectively and having a phase difference of $\pi/2$. In case of light waves, the two superimposed waves must also be coherent, in order to obtain stationary circular or elliptical polarization. This is achieved by deriving the two perpendicularly polarized waves from the same plane polarized wave. The device which is used to split the plane polarized wave also introduces the required phase difference between the two components, and is known as the *phase retardation* or the *wave plate*. The phase retardation plates are made from birefringent crystal cut in a particular manner, and of specified thickness as discussed in the following section.

8.4.1 Phase retardation plates

A phase retardation plate is made from a calcite crystal (or any birefringent crystal) such that the optic axis is parallel to the refracting plane of the plate, and parallel to the edge as shown in Figure

8.17. The thickness of the plate depends on the two refractive indices of the crystal for the wavelength for which it is to be used.

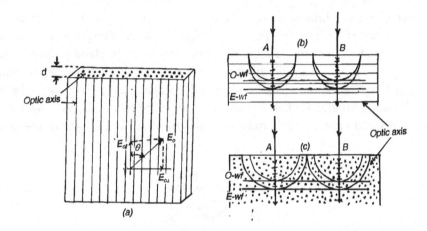

Figure 8.17 (a) Orientation of optic axis in a phase retardation plate, and decomposition of the electric field vector of a plane polarized light incident on the plate Two views of the propagation of the two components through the plate with the optic axis (b) parallel, and (c) perpendicular to the plane of incidence.

Consider plane polarized light with electric field vector of amplitude E_o incident normally on the plate with E_o at an angle θ with the optic axis. The incident E_o-vector is decomposed into components which are parallel and perpendicular to the optic axis components as:

$$E_{o,\parallel} = E_o cos\theta, \quad and \quad E_{o,\perp} = E_o sin\theta \tag{8.9}$$

On entering the plate, the parallel component behaves as an O-ray, and the perpendicular component behaves as an E-ray. Both components travel along the same perpendicular direction through the plate but with the different velocities corresponding to the O- and E-rays respectively (Figure 8.17 b, c). Consequently, as the rays two rays pass through a plate of thickness d, an optical path difference $\Delta x = d \times (n_o - n_e)$ is introduced between them, and they emerge from the crystal with a phase difference ϕ:

$$\phi = \frac{2\pi}{\lambda}\Delta x = \frac{2\pi}{\lambda}d(n_o - n_e) \tag{8.10}$$

where λ is the wavelength of the light. In polarization studies and applications commonly used wave plates are the quarter and half wave plates which introduce a phase difference of a quarter and half a wave respectively. Their thicknesses are given by:

$$d \times (n_o - n_e) = (m + \frac{1}{4})\lambda, \quad (quarter\ waveplate) \tag{8.11}$$

$$d \times (n_o - n_e) = (m + \frac{1}{2})\lambda, \quad (half\ waveplate)$$

where $m = 0, 1, 2, 3, ...(an\ integer)$. The plates have the least thickness for $m = 0$. A non-zero integer value of m simply introduces a phase difference equal to m number of full waves which is equivalent to no phase difference. Thus the integer m in equation (8.11) simply adds m number of full wave plates of thickness $\lambda/(n_o - n_e)$ each to the quarter and half wave plates of the least thickness. The thickness of the wave plates are very small, of the order of the wavelength of light. They are, therefore, sandwiched between glass plates for protection and easy handling.

8.4.2 Elliptical and circular polarization

To produce these polarizations, one uses a quarter wave plate that introduces a phase difference of $\pi/2$ between the O- and the E- waves. The electric field vectors of the O- and E- waves at time t at location x after passing through the quarter wave plate ($\phi = \pi/2$) from eqs. (8.2) and (8.9) are given by:

$$E_y = E_o \cos\theta \sin(\omega t - kx) = E_{o\|} \sin(\omega t - kx) \tag{8.12}$$
$$E_z = E_o \sin\theta \sin(\omega t - kx + \frac{\pi}{2}) = E_{o\perp} \cos(\omega t - kx)$$

from equation (8.12) we get the resultant wave motion of the elliptically polarized light as:

$$\frac{E_y^2}{E_{o\|}^2} + \frac{E_z^2}{E_{o\perp}^2} = 1 \tag{8.13}$$

with $E_{o\|}$ and $E_{o\perp}$ are the two axes of the ellipse.

If the angle of incidence of the field vector of the plane polarized light with the optic axis is $\theta = 45^o$, then $E_{o\|} = E_{o\perp} = E_o cos45$, and the equation of the resultant circular polarized wave is:

$$E_y^2 + E_z^2 = \frac{E_o^2}{2} \tag{8.14}$$

8.5 Analysis and Detection of Polarized Light

The detection of plane polarized light has already been dealt with informally in the earlier sections. When a plane polarized light is viewed through another polarizer, called the analyzer in this mode of application, and rotated about the direction of propagation of the light, intensity of light varies between a maximum for two diagonally opposite positions of the analyzer and a minimum for two in between perpendicular positions. When the intensity is maximum, the analyzer is parallel to the polarizer, and when the intensity is minimum, the analyzer is in the crossed position. In this section, we first deal with the variation of intensity which is given by Malus Law, and then discuss the detection of various types of polarized light.

8.5.1 Malus law

Let E_o be the amplitude of the electric field vector of the plane polarized light, which lies in the plane of transmission of polarizer. Let the plane of transmission of the analyzer makes an angle θ

with that of the polarizer. E_o of the plane polarized light can then be resolved into two components: $E_o cos\,\theta$ in the plane of transmission of the analyzer, and $E_o sin\,\theta$ perpendicular to it (Figure 8.18). Only the parallel component $E_o cos\,\theta$ is transmitted through the analyzer, and the intensity of the transmitted light, which is proportional to the amplitude is:

$$I \propto E_o^2 cos^2\theta = I_o cos^2\theta \qquad (8.15)$$

where I_o is the intensity of the plane polarized light. This is Malus Law. $I = I_o$, for $\theta = 0, 180^o$ and $I = 0$ for $\theta = 90^o, 270^o$. Thus for one complete rotation of the analyzer the intensity of the transmitted plane polarized light passes through a maximum and a minimum alternately, twice each at 90° intervals.

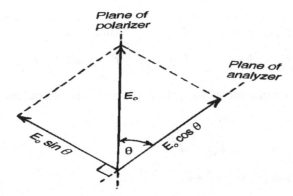

Figure 8.18: Components of plane polarized light incident at an angle θ with plane of transmission of an analyzer.

8.5.2 Detection of variously polarized light

The various possible cases of polarized light are:

- Unpolarized

- Plane polarized

- Partly plane polarized

- Circularly polarized

- Elliptical polarized

Here we consider only the pure case of each polarization. Each one of these can be detected by using an analyzer, such as a Nicol prism, and a quarter wave plate in a particular sequence as follows.

Step - 1 View the beam of light through a rotating Nicol prism.

- Intensity of the light remains constant throughout the complete rotation. The beam is either *unpolarized* or *circularly* polarized.

- Intensity varies between two maxima and two non-zero minima in each rotation. The beam is *partially polarized* or *elliptically* polarized.

- Intensity varies between two maxima and two minima of zero intensity in each rotation. The beam is plane polarized.

Step - 2: If in step-1 light is found to be either unpolarized or circularly polarized.

First pass the beam through a quarter wave plate (held at any orientation), and then view it through a rotating Nicol prism.

- Intensity of the light remains constant. The beam is *unpolarized*.

- Intensity varies between two maxima and two minima of zero intensity in each rotation. The beam is circularly polarized.

Step - 3: If in step-1 light is found to be either partially polarized or elliptically polarized.

First pass the beam through a quarter wave plate whose optic axis is aligned parallel to the direction along which the maximum (or minimum) intensity in step-1 was observed, and then view it through a rotating Nicol prism.

- Intensity varies between two maxima and two minima of zero intensity in each rotation. The beam is elliptically polarized.

- Intensity varies between two maxima and two non-zero minimum in each rotation. The beam is *partially polarized*.

It is left as an exercise for students to explain each observation and conclusions drawn in the above analysis of various forms of polarized light.

8.6 Applications of Polarization

In section 8.3.5 we discussed polarization in nature. The blue colour of the sky and red colour of setting/ rising sun are linked to polarization. Light reflected from smooth surfaces such as the road surface, and from water bodies is partially plane polarized, and the polarization is governed by the principle of polarization by reflection. The polarized component of such reflected light is in the horizontal plane, which causes a glare. The glare can be cut by using Polaroid sunglasses with vertical transmission axis to filter out the horizontal polarized component. The Polaroid sunglasses are a practical, inexpensive device for the comfort and protection of the eye.

Amongst technical applications, some materials exhibit optical activity with polarized light, which depends on the material, concentration if in a solution, wavelength of light, and temperature. By measuring optical activity of the material with plane polarized light concentration of solutions and purity of materials can be determined to reasonable accuracy without having to measure it analytically. Another area of the application of polarized light is the analysis of stress in materials.

8.6.1   Optical activity

When plane polarized light passes through certain crystals, liquids and solutions, the plane of polarization of the emergent beam is rotated. This optical property of substances is known as the *optical activity*. The examples of optically active materials are, quartz, aqueous solution of cane sugar, ice. Looking in to the transmitted beam, the plane of polarization is rotated either clockwise which is called *right-handed* rotation, or it is rotated anti clockwise which is called *left-handed* rotation. The direction of rotation is a characteristic of the material whereas the magnitude of rotation depends on the concentration of the substance if in solution form, length traveled by polarized light in the substance, wavelength of light, and temperature.

The optical activity is expressed in terms of *specific rotation*, $[\alpha]_\lambda^t$ where t is the temperature, and λ is the wavelength. For various substances it is defined as follows:

(i) For Solutions: If m is the solute mass in *g per cc* of the solution, θ is the rotation produced by l cm column of the solution, then the specific rotation of the solution is the rotation produced by 10 cm column of the solution, given as:

$$[\alpha]_\lambda^t = \frac{10\,\theta}{m\,l} \quad (degrees\ per\ (g\ cm^{-3})) \tag{8.16}$$

(ii) For pure liquids: If ρ is the density of the liquid in $g\ cm^{-3}$, θ is the rotation produced by l cm of the liquid column, then the specific rotation of the liquid is the rotation produced by its 10 cm column, given as:

$$[\alpha]_\lambda^t = \frac{10\,\theta}{\rho\,l} \quad (degrees\ per\ (g\ cm^{-3})) \tag{8.17}$$

(iii) For solids such as quartz and ice: If θ is the rotation produced by l cm of length of the solid, then its specific rotation is the rotation produced by 1 mm length of the solid, given as:

$$[\alpha]_\lambda^t = \frac{\theta}{10\,l} \quad (degrees\ per\ mm) \tag{8.18}$$

The optical activity is measured using a polarimeter, that consist of a polarizer, and an analyzer, separated by a space in which the specimen under investigation can be inserted; enclosed in a darkened tube. The liquids are contained in a tube closed with optical glass at both ends. A simple polarimeter used to measure the optical activity of liquids is shown in Figure 8.19. First one determines the direction of the plane of polarization without the sample in place by turning the analyzer till the field of view turns fully dark. Next the sample is inserted between the polarizer and the analyzer. When the polarized light passes through it, its plane of polarization gets rotated clockwise or anti clockwise. The new orientation of the plane of polarization is determined by turning the analyzer again to zero intensity. In order to determine whether the rotation of the plane of polarization is left handed or right handed, two samples of different length must be used. Why? Also explain why it is desirable to make measurements with the dark field of view rather than with maximum intensity orientation of the analyzer?

The optical activity measurement are used in chemical industry to determine the concentration of solutions of optically active substances without having to measure it by more cumbersome analytical means. In the brewing industry, various stages of fermentation are determined by measuring the optical activity of the sugar content in the brew such as alcohol or wine.

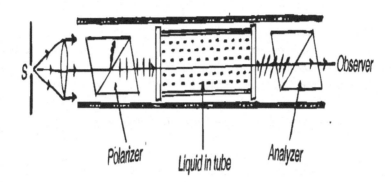

Figure 8.19: A simple polarimeter to measure the optical activity of liquids.

8.6.2 Optical stress analysis

Optical stress analysis using plane polarized light is an important industrial application for non-destructive testing. When a specimen of carefully annealed glass or plastic material is placed between two crossed Niclos, the field of view remains dark if the specimen is not double refracting. If the specimen is doubly refracting the field of view turn into partially bright. Certain glasses and plastics become doubly refracting under stress. The pattern of interference fringes reveals the portion of the material under stress and hence the weak and strong points in the sample. Non-transparent objects are studied with reflected light.

8.7 Chapter 8 Summary

The Chapter deals with the production and detection of polarized light and its characteristics. Polarization is possible only for the transverse waves, (displacement vector perpendicular to the direction of propagation of the wave). Thus, polarization is the only and conclusive evidence of the transverse nature of light. The magnetic field vector **B** of EM wave being related to the electric filed vector **E**, we shall discuss polarization of light only in terms of the **E** vector.

- In an un-polarized EM wave, the oscillatory **E** vector can be in any possible transverse direction, but in a plane polarized wave the **E** vector lies only in one plane containing the direction of wave propagation, known as the plane of polarization. In addition there are circular and elliptical polarized light.

- A device used for the production of plane polarized light is known as Polarizer. The same device can also be used for the detection of the polarization of light in which case it is known as the analyzer.

- Circular and Elliptical polarized light are obtained from the superposition of two perpendicular, plane polarized waves of the same frequency. The nature of the resulting wave depends on the phase difference f between the two waves and their amplitudes as follows:

  ○ If $\phi = 0$ or π, the resulting wave is plane polarized with its plane of polarization rotated by $\pm 45°$.

  ○ If $\phi = \pi/4$ or $3\pi/4$, the resulting wave is elliptically polarized.

  ○ If $\phi = \pi/2$ the resulting wave is elliptically polarized if the amplitudes of the two waves are un-equal, and circularly polarized if their amplitudes are equal.

  ○ Elliptically polarized lights for $\phi = \pi/4, 3\pi/4$ and $\pi/2$ differ in the orientation of the major axis of the ellipse.

  ○ Note that circle is a special case of an ellipse with equal major and minor axes.

- In a circular polarized light, the constant **E** vector, equal to that of individual waves rotates in a circle about the direction of propagation of the wave with frequency ?, equal to that of individual waves. Looking into the direction of the oncoming wave gives two types of circular polarized light:

  ○ If the **E** vector rotates counter-clockwise, the light is left circularly polarized and has positive helicity.

  ○ If the **E** vector rotates clockwise, the light is right circularly polarized and has negative helicity.

- For elliptically polarized light the rotating **E** vector varies between the amplitudes of the two superimposed plane polarized waves which define the semi-major and semi-minor axes of the ellipse. Both, counterclockwise and clockwise rotations are possible as in circular polarization.

- Plane polarized light can be produced by the following methods:

  ○ Reflection: Light reflected from a dielectric medium (glass plate) at Brewsters angle, $\theta_p = \tan^{-1}(n)$ is 100% plane polarized with **E** vector perpendicular to the plane of incidence. At any other angle it is partly polarized.

  ○ Refraction: As the light with **E** vector perpendicular to the plane of incidence is removed on reflection, the transmitted light is partly plane polarized with **E** in the plane of incidence. By using a stack of glass plates, transmitted light can also be made 100% plane polarized.

  ○ Double refraction: When un-polarized light is refracted through a crystal of Calcite, NaNo3, Quartz and some other materials, the crystal is optically homogeneous to light with **E** perpendicular to the plane of incidence (o-wave) and inhomogeneous to light with **E** in the plane of incidence (E-wave). Along the optic axis both o- and E waves have the same refractive index, and travel together. This property of these crystals is known as Double refraction. A Nicol Prism (not a prism in geometrical sense) made of any of these materials is used to produce plane polarized light. Geometrical prisms of these materials with specific orientations of the optic axis are used to measure the refractive indices of the material for o- and E-waves by minimum deviation method.

  ○ Selective absorption: Crystals of some doubly refracting materials, such as Tourmaline completely absorb plane polarized light with E in one direction and transmit plane polarized light with **E** in the perpendicular direction.

  ○ Polaroid sheets: These are the large area, commercially produced plastic sheets with dichroic

properties that are used for the production of plane polarized light. They contain parallel oriented long hydrocarbon chains attached to iodine atoms that give the polarizing property to the sheets.

○ Scattering: Light passing through suspended fine dust particles undergoes scattering and the scattered light is plane polarized. The sky-light that gives it the blue colour is plane polarized with planes of polarization randomly oriented.

○ Circularly and elliptically polarized light are produced using a quarter wave plate made from a double refracting crystal with optic axis parallel to the edge of the plate. When plane polarized light passes through the plate it emerges as circularly or elliptically polarized light depending of the orientation of the incident **E** vector.

• Detection of variously polarized light: A polarizer is also used as an analyzer of polarized light. When the polarizer and the analyzer are parallel, the plane polarized light from the polarizer passes through the analyzer, and when they are cross, the polarized light from the polarizer does not pass through the analyzer. At any other angle, Malu's Law: $I = I_o \cos \theta$ applies. For the step wise procedure for the analysis of plane, circular and elliptical polarized lights, readers are referred to section 8.5.2 of the chapter.

• Optical activity and application of polarization of light: When plane polarized light passed through certain crystals, liquids, and solutions, the plane of polarization is rotated. The rotation depends on the length of light passage through the medium and the concentration of solutions. This property of materials is known as the optical activity. It is used in chemical and brewing industries to determine the stage of the process through concentration determination from optical activity measurement without having to carry out sometimes complex chemical analysis.

8.8 Exercises

8.1 Calculate the maximum and minimum intensities of the beam for the two perpendicular positions of the analyzer in terms of the total intensity of the partially polarized incident beam, if the polarization is 30%.

8.2 Refractive indices of certain sample of glass and water are 1.51, and 1.33 respectively. Calculate Brewster's angle (i) from air to glass, (ii) from air to water, and (iii) from water to glass. Calculate the angle of the refracted ray in each case, without applying Snell's law. Is it feasible to have Brewster's angle for reflection from (iv) glass to air and (v) water to air surfaces? Discuss.

8.3 Sun light reflected from a lake is pure plane polarized. Calculate the position of the sun above horizon.

8.4 Light reflected from a specimen of glass is completely plane polarized for an angle of incidence $58°23'$. Calculate the refractive index of the specimen.

8.5 Using Brewster's relationship, prove that at the Brewster's angle of incidence the reflected and the refracted rays are perpendicular to each other.

8.6 Consider light of wavelength λ_1 and λ_2, with Brewster's angles θ_1 and θ_2 respectively. Express θ_2 in terms of λ_1, λ_2 and θ_1, using the simple refractive index and wavelength relationship.

8.7 Calculate the polarization of a beam by transmission through a pile of 15 glass plates, each of refractive index 1.47. What is the angle of incident of the beam at the pile of plates?

8.8 Draw a labeled diagram to show the refraction of a beam of light through a quartz crystal following Huygen's wave theory for the following cases of the orientation of the optic axis. Optic axis is *(i)* perpendicular to the crystal surfaces, *(ii)* parallel to the crystal surface and in the plane of incidence, and *(iii)* parallel to the crystal surface and perpendicular to the plane of incidence.

8.9 Calculate the angle of total internal reflection for both the O- and the E- rays at the calcite crystal-canada balsam interface. Explain qualitatively why the O-ray is internally reflected and not the E-ray.

8.10 For which doubly refracting crystals listed in Table 8.1 canada balsam can be used as a cement to eliminate one of the plane polarized beams by internal reflection. Calculate the critical angles for both the rays at the crystal-canada balsam interface for each crystal. If a prism similar to Nicol prism is made from these crystals, which of the two beams shall be eliminated in each case.

8.11 Calculate the angles of minimum deviation for O- and E- rays for the double refracting crystals listed in Table 8.1. The angle of the prism is 60°.

8.12 Calculate the least thickness for quarter, half and full wave plates for the sodium yellow light for the materials listed in Table 8.1. Assuming that the refractive index of these materials depended on wavelength, the phase difference introduced by each of the plates for the two extreme wavelengths, red and violet, of the visible spectrum. Tabulate your results appropriately, and what general conclusions can you draw for the thickness of wave plates in relation to wavelength of light.

8.13 Light reflected from diamond at 57° $32.4'$ when viewed with a rotating Nicol has maximum and zero intensity for two perpendicular orientations respectively. Calculate the refractive index of diamond.

8.14 Two mutually perpendicular plane polarized beams *A* and *B* traveling parallel to each other are examined with a rotating Nicol prism. At certain orientation of the Nicol, the intensity of beam *A* is maximum, while that of *B* is zero. When the Nicol is further rotated by 35°, the intensity of both the beams are found to be equal. Calculate the ratio of the intensities of the two beams. For what orientation of the Nicol relative to the very first orientation, the intensity of beam *A* will be *(i)* twice, and *(ii)* half of the intensity of beam *B*.

8.15 What fraction of the intensity of plane polarized light is transmitted through a Polaroid sheet if its axis of transmission makes an angle of 37° with the axis of transmission of the polarizer.

8.16 Two Nicol prisms used as a polarizer and analyzer are set in a crossed position. A Polaroid sheet is inserted between the two, and it is rotated about the beam from the polarizer. Starting with the axis of transmission of the Polaroid parallel to that of the polarizer, calculate the intensity of the light transmitted through the analyzer as a function of the angle of rotation θ of the Polaroid sheet.

8.17 Rotation of the plane of polarization by a sample of wine contained in a 12 cm tube of a polarimeter is 121°. If the specific rotation of sugar is $50^o per\ (gcm^{-3})$, calculate the sugar concentration in the wine.

8.18 A plane polarized beam is viewed through a rotating Polaroid, and one of the maxima of intensity is observed at $\theta = 12^o$. Calculate the ratio I/I_{max} for the angular orientations $\theta = 42^o$ and 140° of the Polaroid.

8.19 Explain why unpolarized light can not be used to produce elliptical and circular polarized light using a quarter wave plate.

8.20 What happens to unpolarized light, and plane polarized light when they pass through a quarter wave plate?

8.21 You are given a glass plate. Using two Nicols how would you find out whether the given plate is a quarter wave plate, or a half wave plate or an ordinary glass plate.

8.22 A plane polarized light is passed through a half wave plate. Describe the polarization of the emergent light for various orientations of the optic axis of the wave plate.

8.23 A quartz plate has a thickness of 0.1234 mm. For what wavelengths in the visible region $(\lambda = 450 - 800nm)$ it will act as a *(i))* quarter wave plate, and *(ii)* half wave plate? Ignore the dependence of refractive index on wavelength.

Chapter 9

Quantum Optics

9.1 Introduction

Quantum optics is a study of light which makes use of the particle-like property of light. There is a lot of experimental evidence which confirms the particle nature of light as was discussed in Chapter 1, section 1.4.3, where it was discussed that a photon is a quantum of light with energy, $E = h\nu$. The interaction of light with an atom is central to understanding quantum optics, and for this reason we shall first study the structure of the atom in this chapter. In later sections of the chapter we shall study topics of absorption, emission (spontaneous and stimulated), scattering, lasers, holograms.

9.2 Bohr's theory of the atom

In *Bohr's theory of the atom*, the following basic assumptions are made:

(i) In an atom, an electron of mass m_e moves with a velocity v around a nucleus in orbits of radii r such that the angular momentum L of the electron is *quantised* in multiples of $\hbar$. While in these orbits, the electron does not radiate.

$$L = m_e vr = n\hbar \text{ where } n = 1, 2, 3, \cdots \tag{9.1}$$

and hence

$$r = \frac{n\hbar}{m_e v} \tag{9.2}$$

(ii) When an electron makes a transition from a higher energy level orbit to a lower energy level orbit, it is accompanied by *emission* of radiation, and when an electron *absorbs* a photon it moves from a lower energy level to a higher energy level. The change in energy between the initial energy level E_i and the final energy level E_f is related to the frequency ν of the associated photon as:

$$E_i - E_f = h\nu \tag{9.3}$$

Although the dynamics of an electron obey quantum mechanical laws, it is possible to get a reasonable understanding if we approximate the dynamics using a classical equation of motion (from Newton's second law) as below. The electron moving in a circular orbit experiences an acceleration $a = v^2/r$. The required centripetal force for the circular motion is supplied by the coulomb interaction between the electron and the nucleus. Thus we have:

$$m_e a = \sum F$$

$$m_e \frac{v^2}{r} = \frac{kZe^2}{r^2}$$

$$\text{or} \quad v^2 = \frac{Ze^2}{4\pi\epsilon_0 m_e r} \tag{9.4}$$

where Z is the atomic number of the atom, $k = 1/(4\pi\epsilon_0)$ is the Coulomb's force constant, and $\epsilon_0 = 8.854 \times 10^{-12} c^2 N^{-1} m^{-2}$ is the permittivity of free space. From Bohr's theory, the radius r_n for an electron in an orbit n can be obtained as below.

$$v = \frac{n\hbar}{m_e r}$$

Using equations (9.2) and (9.4),

$$\frac{Ze^2}{4\pi\epsilon_0 m_e r} = \frac{n^2\hbar^2}{m_e^2 r^2}$$

$$\tag{9.5}$$

and hence the radius of the nth orbit is given as

$$r_n = \frac{4\pi\epsilon_0 n^2\hbar^2}{m_e Ze^2} \tag{9.6}$$

where for $n = 1$ and $Z = 1$, we have

$$r_1(\text{hydrogen}) = \frac{4\pi\epsilon_0 \hbar^2}{m_e e^2} = a_0 \tag{9.7}$$

known as the *Bohr's radius*, and $r_n = (a_0/Z)n^2$. From equations (9.2) and (9.5), the velocity v_n for an electron in an orbit n can be obtained as below.

$$v_n^2 = \left(\frac{Ze^2}{4\pi\epsilon_0 n\hbar}\right)^2$$

$$v_n = \frac{Ze^2}{4\pi\epsilon_0 n\hbar} \tag{9.8}$$

where for $n = 1$ and $Z = 1$ (for hydrogen atom), we have

$$v_1(\text{hydrogen}) = \frac{e^2}{4\pi\epsilon_0 \hbar} = v_0$$

is the speed of the electron in the first Bohr's orbit of the hydrogen atom, and $v_n = v_0 Z/n$.

Thus the speed of the electron in the nth atomic orbit is inversely proportional to the quantum number n, and decreases in the ratio, $1 : \frac{1}{2} : \frac{1}{3} : \frac{1}{4} : \frac{1}{5} ...$ for $n = 1, 2, 3, 4, 5, ...$.

The total energy E_n for an electron in the nth orbit is the sum of its kinetic energy (KE) and potential energy (PE), and can be obtained as below.

$$
\begin{aligned}
E_n &= KE + PE \\
&= \frac{1}{2}m_e v_n^2 - \frac{1}{4\pi\epsilon_0}\frac{Ze^2}{r_n} \\
&= -\frac{m_e Z^2 e^4}{2\hbar^2(4\pi\epsilon_0)^2}\frac{1}{n^2}
\end{aligned} \tag{9.9}
$$

When the above results are applied to the hydrogen atom $(Z = 1)$, the energy levels are obtained as

$$
\begin{aligned}
E_n &= -\frac{m_e e^4}{2\hbar^2(4\pi\epsilon_0)^2}\frac{1}{n^2} = -E_0\frac{1}{n^2} \\
&= -\frac{(9.11 \times 10^{-31})(1.6 \times 10^{-19})^4(9 \times 10^9)^2}{2(1.05 \times 10^{-34})^2}\frac{1}{n^2} \\
&= -(2.19 \times 10^{-18})\frac{1}{n^2} \quad \text{J} \\
&= -\frac{13.6}{n^2} \quad \text{eV}
\end{aligned} \tag{9.10}
$$

where $E_0 = 13.6$ eV $= 2.19 \times 10^{-18}$ J is the *ionization energy* of the hydrogen atom..

Spectral wavelengths for the hydrogen spectrum can be obtained from the expression of the energy levels of the hydrogen atom. Let E_i and E_f be the initial and final energy levels respectively in a hydrogen atom. Then

$$
E_i = -\frac{13.6}{n_i^2} \quad \text{eV}
$$

$$
E_f = -\frac{13.6}{n_f^2} \quad \text{eV}
$$

$$
E_i - E_f = h\nu = -13.6\left(\frac{1}{n_i^2} - \frac{1}{n_f^2}\right) \times 1.6 \times 10^{-19} \quad \text{J}
$$

$$
\frac{hc}{\lambda} = 13.6\left(\frac{1}{n_f^2} - \frac{1}{n_i^2}\right) \times 1.6 \times 10^{-19}
$$

$$
\frac{1}{\lambda} = 1.097 \times 10^7\left(\frac{1}{n_f^2} - \frac{1}{n_i^2}\right)
$$

$$
\frac{1}{\lambda} = R\left(\frac{1}{n_f^2} - \frac{1}{n_i^2}\right) \quad \text{m}^{-1} \tag{9.11}
$$

where

$$
R = \frac{13.6 \times 1.6 \times 10^{-19}}{hc} \tag{9.12}
$$

$$= 1.097 \times 10^7 \text{ m}^{-1} \tag{9.13}$$

is the Rydberg constant. If $n_i < n_f$ there is absorption, and for $n_i > n_f$ there is emission. An energy level diagram for the hydrogen is shown in Figure 9.1, where the Lyman series, Balmer series, Paschen series are illustrated. The series also includes the Bracket series and Pfund series as defined below.

$$\frac{1}{\lambda} = R\left(\frac{1}{1^2} - \frac{1}{n_i^2}\right) \text{ for Lyman series, } n_f = 1, n_i = 2, 3, 4, \cdots$$

$$\frac{1}{\lambda} = R\left(\frac{1}{2^2} - \frac{1}{n_i^2}\right) \text{ for Balmer series, } n_f = 2, n_i = 3, 4, 5 \cdots$$

$$\frac{1}{\lambda} = R\left(\frac{1}{3^2} - \frac{1}{n_i^2}\right) \text{ for Paschen series, } n_f = 3, n_i = 4, 5, 6 \cdots$$

$$\frac{1}{\lambda} = R\left(\frac{1}{4^2} - \frac{1}{n_i^2}\right) \text{ for Bracket series, } n_f = 4, n_i = 5, 6, 7 \cdots$$

$$\frac{1}{\lambda} = R\left(\frac{1}{5^2} - \frac{1}{n_i^2}\right) \text{ for Pfund series, } n_f = 5, n_i = 6, 7, 8, \cdots$$

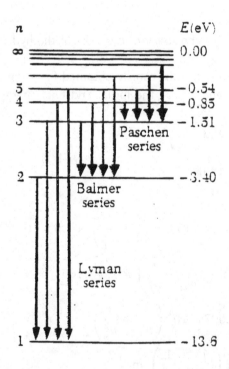

Figure 9.1: An energy level diagram for the hydrogen atom.

9.3 Emission and absorption

Light interacts with an atom in three ways: absorption, spontaneous emission and stimulated emission, as illustrated in Figure 9.2.

In *absorption*, an atom in an energy level E_1 absorbs a photon of energy $h\nu$, and its energy is raised to a higher quantised energy level E_2, such that

$$E_2 - E_1 = h\nu \qquad (9.14)$$

In *spontaneous emission*, an atom at a higher energy level E_2 may on its own accord emit a photon of energy $h\nu$ and and leave the atom in a lower energy level E_1.

In *stimulated emission*, an atom at a higher energy level E_2 may be stimulated by a photon of energy $h\nu$ to emit another photon of the same energy, and finally leave the atom in a lower energy level E_1.

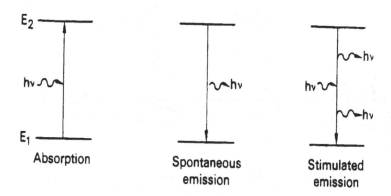

Figure 9.2: Light interacts with an atom in three ways: absorption, spontaneous emission and stimulated emission.

9.3.1   Einstein's A and B coefficients

Einstein's theory of absorption, spontaneous emission and stimulated emision is based on a physically reasonable model, developed in 1917 . Consider a cavity with N atoms, such that N_1 atoms are at a lower energy level with energy E_1 and N_2 atoms are at a higher energy level with energy E_2. Let the total energy density in a closed cavity be $U(\omega)$, consisting of a thermal part $U_T(\omega)$ and external radiation $U_E(\omega)$, *ie*,

$$N_1 + N_2 = N \qquad (9.15)$$

and

$$U(\omega) = U_T(\omega) + U_E(\omega) \tag{9.16}$$

Let A_{21} be the transition rate that the atom will spontaneously fall from state 2 to the lower state 1 and emit a photon of energy $\hbar\omega$, and

$B_{12}U(\omega)$ be the transition rate of an upward transition from state 1 to 2 in presence of radiation $U(\omega)$ due to absorption of a photon of energy $\hbar\omega$, and

$B_{21}U(\omega)$ be the transition rate of a stimulated emission from state 2 to 1 in presence of radiation $U(\omega)$ and a photon of energy $\hbar\omega$ is emitted

These processes are illustrated in Figure 9.3.

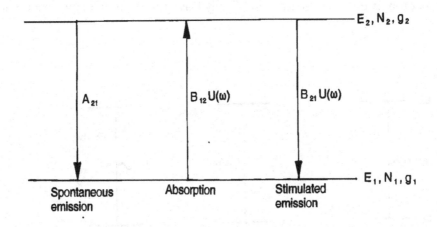

Figure 9.3: Absorption, spontaneous emission and stimulated emission as transitions between two levels of an atom.

The rate equation is

$$\frac{dN_1}{dt} = -\frac{dN_2}{dt} = N_2 A_{21} + N_2 B_{21} U(\omega) - N_1 B_{12} U(\omega) \tag{9.17}$$

For thermal equilibrium

$$\frac{dN_1}{dt} = -\frac{dN_2}{dt} = 0 \tag{9.18}$$

and hence

$$N_2 A_{21} + N_2 B_{21} U(\omega) - N_1 B_{12} U(\omega) = 0 \tag{9.19}$$

which can be rearranged to give

$$U(\omega) = \frac{A_{21}}{(N_1/N_2)B_{12} - B_{21}} \tag{9.20}$$

The ratio N_1/N_2 can be obtained using statistical mechanics, where according to Boltzmann's law, the population levels N_1 and N_2 with degeneracies g_1 and g_2 respectively are given by

$$N_1 \ \propto \ g_1 e^{-E_1/K_B T}$$
$$N_2 \ \propto \ g_1 e^{-E_2/K_B T}$$

which reduces to

$$\frac{N_1}{N_2} \ = \ \frac{g_1}{g_2} e^{(E_2 - E_1)/k_B T}$$
$$\frac{N_1}{N_2} \ = \ \frac{g_1}{g_2} e^{\hbar\omega/k_B T} \tag{9.21}$$

Using equation (9.20) in (9.19), the energy density can be written as

$$U(\omega) = \frac{A_{21}/B_{21}}{(g_1/g_2)(B_{12}/B_{21})e^{\hbar\omega//k_B T} - 1} \tag{9.22}$$

However, according to Planck's radiation law, the energy density which was given in equation (1.21) can also be written in the form

$$U(\omega) = \frac{\hbar\omega^3}{\pi^2 c^3} \frac{1}{[e^{\hbar\omega//k_B T} - 1]} \tag{9.23}$$

The result of the energy density from Einstein's theory given in equation (9.21) and that from Planck's radiation law given in equation (9.22) agree if and only if

$$\frac{A_{21}}{B_{21}} = \frac{\hbar\omega^3}{\pi^2 c^3} \tag{9.24}$$

and

$$\frac{g_1}{g_2}\frac{B_{12}}{B_{21}} = 1 \tag{9.25}$$

and hence equations (9.21) and (9.22) give

$$U(\omega) = \frac{\hbar\omega^3}{\pi^2 c^3}\bar{n} = \frac{A_{21}}{B_{21}}\bar{n} \tag{9.26}$$

where

$$\bar{n} = \frac{1}{[e^{\hbar\omega//k_B T} - 1]} \tag{9.27}$$

From the above discussion we can obtain the following results

$$A_{21} + B_{21}U(\omega) \ = \ A_{21}(\bar{n} + 1) \tag{9.28}$$
$$\frac{A_{21}}{B_{21}U(\omega)} \ = \ \frac{1}{\bar{n}} \tag{9.29}$$

where equation (9.27) gives the *sum* of the two emission rates and equation (9.28) gives the *ratio* of the two emission rates, namely the spontaneous and the stimulated emissions.

9.3.2 Macroscopic theory of absorption

Absorption of electromagnetic waves by a material occurs when the material through which it is propagating has a complex refractive index, $n(\omega)$ and hence a complex dielectric function, $\epsilon(\omega)$. Consider an electric field E given by

$$E(z,t) = E_0 e^{i(kz - \omega t)} \tag{9.30}$$

The wavevector k satisfies

$$\frac{c^2 k^2}{\omega^2} = \epsilon(\omega) = \epsilon'(\omega) + i\epsilon''(\omega) \tag{9.31}$$

where $\epsilon'(\omega)$ and $\epsilon''(\omega)$ are the real and imaginary parts of the dielectric function. Also, we have

$$\frac{ck}{\omega} = n(\omega) = \eta(\omega) + i\kappa(\omega) \tag{9.32}$$

where $\eta(\omega)$ is the real part of the refractive index and $\kappa(\omega)$ is the imaginary part, usually called the extinction coefficient. The refractive index is the square root of the dielectric function.

$$n(\omega) = \sqrt{\epsilon(\omega)} \tag{9.33}$$

or

$$n^2(\omega) = \epsilon(\omega)$$

From equations (9.30) and (9.31), we obtain

$$[\eta(\omega) + i\kappa(\omega)]^2 = [\epsilon'(\omega) + i\epsilon''(\omega)]$$

which can be cast in the form

$$\eta^2(\omega) - \kappa^2(\omega) + i2\eta(\omega)\kappa(\omega) = \epsilon'(\omega) + i\epsilon''(\omega)$$

Separating the real and imaginary parts, we have

$$\epsilon'(\omega) = \eta^2(\omega) - \kappa^2(\omega) \tag{9.34}$$

$$\epsilon''(\omega) = 2\eta(\omega)\kappa(\omega) \tag{9.35}$$

With a little algebra (see exercise 9.2), equations (9.33) and (9.34) can be cast as quadratic equations for $\eta^2(\omega)$ and $\kappa^2(\omega)$ in the form

$$4\eta^4(\omega) - 4\epsilon'(\omega)\eta^2(\omega) - \epsilon''(\omega)^2 = 0$$

and

$$4\kappa^4(\omega) + 4\epsilon'(\omega)\kappa^2(\omega) - \epsilon''(\omega)^2 = 0$$

The above quadratic equations have the following solutions

$$\eta(\omega) = \left\{ \frac{1}{2}[\epsilon'(\omega) + \sqrt{(\epsilon'(\omega)^2 + \epsilon''(\omega)^2)}] \right\}^{1/2} \tag{9.36}$$

$$\kappa(\omega) = \{\frac{1}{2}[-\epsilon'(\omega) + \sqrt{(\epsilon'(\omega)^2 + \epsilon''(\omega)^2)}]\}^{1/2} \tag{9.37}$$

Thus, the propagating wave, from equations (9.29), (9.31), (9.35) and (9.36) is of the form

$$E(z,t) = E_0 e^{-\frac{\omega}{c}\kappa z} e^{i(\frac{\omega}{c}\eta z - \omega t)} \tag{9.38}$$

which shows that the amplitude will be decaying exponentially due to absorption. The quantities $\eta(\omega)$, $\kappa(\omega)$, or equivalently $\epsilon'(\omega)$, $\epsilon''(\omega)$ are important in studying optical properties of materials, and equations (9.35) and (9.36) are applied in such studies.

9.4 Scattering of light

Scattering of light can either be *elastic* or *inelastic*. In elastic scattering, there is no change in the frequency of the incident light, while in inelastic scattering there is a shift in the frequency. An example of elastic scattering is Rayleigh scattering, while examples of inelastic scattering are Raman scattering by optical phonons in solids and Brillouin scattering by acoustic phonons in solids.

In this section, we consider a simple example of elastic scattering, known as Rayleigh scattering. The differential cross section is given by

$$\frac{d\sigma}{d\Omega} = \left(\frac{e^2}{4\pi\epsilon_0 mc^2}\right)^2 \frac{\omega^4}{(\omega_0^2 - \omega^2)^2 + \omega^2\Gamma^2}(\vec{\epsilon}.\vec{\epsilon_s})^2 \tag{9.39}$$

where the factor

$$\frac{e^2}{4\pi\epsilon_0 mc^2} = r_e \tag{9.40}$$

is the classical electron radius and has the magnitude 2.8×10^{-15} m. The vectors $\vec{\epsilon}$ and $\vec{\epsilon_s}$ are unit polarisation vectors in the direction of incident radiation and scattered radiation respectively. The frequency dependence of the differential cross-section for Rayleigh scattering is illustrated in Figure 9.4.

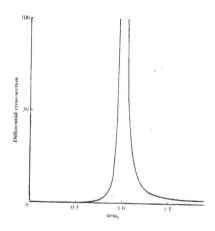

Figure 9.4: The frequency dependence of the differential cross-section for Rayleigh scattering.

It can be noted that there is a large peak that occurs at the resonant frequency ω_0. At low frequencies, the cross-section increases with the fourth power of frequency, and it is this dependence that accounts for the blue colour of the sky and the red of sunset. At high frequencies, the differential cross-section tends to a constant value which is proportional to the classical electron radius, r_e, as can be seen in Figure 9.4.

9.5   Lasers

The word LASER is an acronym for Light Amplification by Stimulated Emision of Radiation. A laser is a light source, but very different from a conventional light source. Light from a laser source is:

(i) Monochromatic
(ii) Coherent
(iii) narrow and highly collimated
(iv) Parallel
(v) very intense

In a laser, atoms or molecules of a gas (such as He-Ne, Argon, CO_2, N_2), liquid or solid (such as ruby, neodymium/yttrium aluminium garnet (Nd:YAG)) are excited in a *laser cavity*, of say length L. The laser cavity has reflecting surfaces at its ends so that photons can be reflected back and forth. One of the reflecting surfaces must be partially transparent so that light can escape the cavity and be generated as a laser beam which can be used for several applications. A standing electromagnetic wave fills the space between these two mirrors. This wave must satisfy the boundary condition that its electric field is zero at the position of the mirrors. The condition for the standing waves to be formed is that the distance L between the mirrors and the wavelength λ of the wave are related by

$$L = \left(\frac{n+1}{2}\right)\lambda \qquad\qquad n = 0,1,2,\cdots$$

Within the laser cavity, conditions are created such that there is *population inversion*. If N_1 and N_2 are the numbers of atoms at energy levels E_1 and E_2, then according to statistical mechanics, at a temperature T the ratio is given by

$$N_2 = N_1 e^{-(E_2-E_1)/K_B T} \tag{9.41}$$

where population inversion occurs if $N_2 > N_1$, that is more atoms are at the higher energy level than there are at the lower energy level.

Wavelengths of some typical lasers are shown in Table 9.1 . Typical laser systems such as the He-Ne laser and Ruby laser are illustrated in Figures 9.5, 9.6 and 9.7 respectively.

Table 9.1: Wavelengths of some tpical lasers.

Gas lasers	
He-Ne	$6328\overset{\circ}{A}$
Argon ion	$4880\overset{\circ}{A}$, $5145\overset{\circ}{A}$
N_2	$3370\overset{\circ}{A}$
CO_2	9.6 to 10.6μm
Solid state lasers	
Ruby(Cr^{3+})	$6943\overset{\circ}{A}$
Nd/YAG	1.06μm plus other near infrared lines
GaAs	$9000\overset{\circ}{A}$
GaAsP	$6500\overset{\circ}{A}$
Dye lasers	
Rhodamine G	$5710\overset{\circ}{A}$

Applications of lasers include

(i) Industrial applications such as cutting and boring metals
(ii) Medical applications such as surgery
(iii) Scientific research in biology, chemistry, physics and other branches
(iv) Communications
(v) Holography
(vi) Technological devices such as laser printers, CD players, scanners, bar-code scanners used in supermarkets and many others.

He-Ne Laser:

The He-Ne laser is illustrated in Figure 9.5 and the relevant simplified energy level diagram is shown in Figure 9.6, where it can be noted that levels 2 of He and Ne are approximately the same. Laser action occurs between levels 2 and 3 of Ne. It should be noted that level 2 is not excited by photons but rather by collisions with He atoms excited by an electric discharge in the He-Ne mixture.

Ruby Laser:

The ruby laser is illustrated in Figure 9.7 and the associated simplified energy level diagram is shown in Figure 9.8. Ruby consists of small concentrations of Cr^{3+} ions in a lattice of Al_2O_3, and the relevant simplified energy level diagrams are shown in Figure 9.8. Cr^{3+} ions are excited from the ground state (level 1) to level 2 by absorption of photons from a xenon flash tube which surrounds the ruby rod. Many of the ions decay from level 2 to level 3 (a metastable state). Thus after a short time, the population of level 3 may become greater than that of level 1, thus enabling "population inversion" to be achieved. With the help of mirrors at the ends, multiple reflections occur and the laser action (light amplification) takes place.

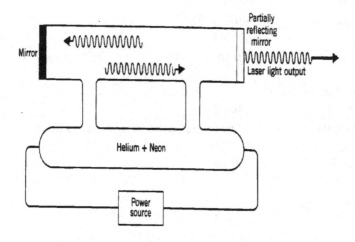

Figure 9.5: Basic structure of the He-Ne Laser.

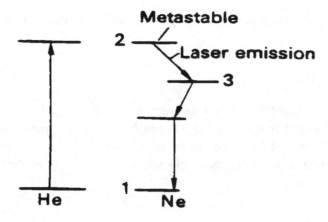

Figure 9.6: Simplified energy level diagram for the He-Ne Laser.

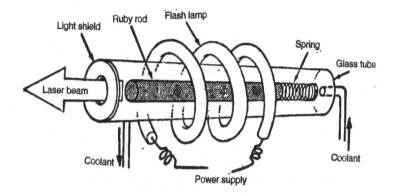

Figure 9.7: Basic structure of the Ruby Laser.

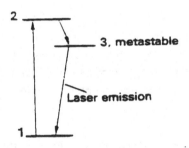

Figure 9.8: Simplified energy level diagram for the Ruby Laser.

9.6 Holography

Holography is a technique for producing a *hologram* that gives real life like 3-D images in space. The principles of holography were first enunciated by Dennis Gabor in 1948, and he was awarded a Nobel prize for his work in 1971. The difference between an ordinary photograph and a hologram is that in a photograph only the intensity distribution received from the object is recorded, while in a hologram both the intensity and the phase of light received from the object are recorded. Hence a hologram shows the three-dimensional nature of the object. A hologram is produced using a laser, and involves two processes, as illustrated in Figure 9.9.

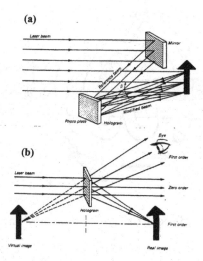

Figure 9.9: (a) Illustration of how a hologram is produced. (b) Illustration of how a hologram is viewed.

(i) An incident laser beam is split into two beams as illustrated in Figure 9.9 (a). One of the beams is due to reflection from a plane mirror, and the reflected beam is referred to as the *Reference beam*. The other beam is scattered by the object is referred to as the *Modified beam*. The two beams interfere and the resulting interference pattern known as a hologram can be captured on a photosensitive plate such as a film.

(ii) To view the hologram, the interference pattern on the film is illuminated by a laser beam of the same frequency, and a three-dimensional image of the original object appears, as illustrated in Figure 9.9 (b).

9.7   Chapter 9 Summary

Quantum optics deals with optical processes and phenomenon which can only be understood in terms of the quantum nature of light, the photon of energy $E = h\nu$. The notable ones are the emission and absorption of light by atoms and molecules. The basis of these processes is the structure of the atom. This Chapter presents the simplest model of the atom, the Bohrs model, which is then applied to explain some of the quantum optical processes and phenomenon.

- Salient features of the Bohrs atomic model are:
  - Electrons in the atom orbit the nucleus in circular orbits of radius r such that their angular momentum L is quantized in the units of $h : L = n\hbar, n = 1, 2, 3, \cdots (\hbar = h/2\pi)$
  - orbital energy of electron is quantized, given as: $E_n = (-)13.6/n^2$ (for the hydrogen atom). The negative sign signifies the bound state of the electron in the atom.

○ The orbital radius is quantized, given as: $r_n = r_1 n^2$, where $r_1 = \frac{4\pi\epsilon_0 \hbar^2}{m_e e^2}$.

○ The orbital velocity of the electron atom is quantized as: $v_n = v_1/n$, where $v_1 = \frac{e^2}{4\pi\epsilon_0 \hbar}$.

○ $n = 1$ gives the ground (stable) state of the atom. Higher values of n give the unstable excited states of the atom.

○ For $n \to \infty$, E, and $v \to 0$, and $r \to \infty$. This represents the ionized state of the atom with e far removed from the atom to a free state. ○ When an atom absorbs a photon of appropriate energy, the electron moves to a higher orbit and the atom is in excited state, where it can stay only for extremely short duration of time. This results in atomic absorption spectra.

○ An atom in excited state returns to ground state by emitting a photon/ photons of appropriate energy/ energies resulting in characteristic atomic spectral lines.

- Spectral series in the emission spectra of hydrogen atom are defined as:
  ○ Lyman series: $n_f = 1, n_i = 2, 3, 4, \cdots$
  ○ Balmer series: $n_f = 2, n_i = 3, 4, 5, \cdots$
  ○ Paschen series: $n_f = 3, n_i = 4, 5, 6, \cdots$
  ○ Bracket series: $n_f = 4, n_i = 5, 6, 7, \ldots$
  ○ Pfund series: $n_f = 5, n_i = 6, 7, 8, \cdots$

- An atom having absorbed a photon can undergo either spontaneous emission or stimulated emission, rates of which are expressed in terms of the Einsteins coefficients A and B.

- Photons can also undergo elastic or inelastic scattering from atoms, theory of which also falls under quantum optics.

- LASER is an acronym for Light Amplification by Stimulated Emission of Radiation. Lasers produce monochromatic, coherent, highly collimated, parallel beam of high power density. To produce a laser beam, the lasing atoms are first excited using an external energy source to a higher energy state in large numbers, known as the inversion of state. Then they are made to undergo stimulated emission simultaneously resulting in a laser beam. Notable lasers in common use in teaching laboratories are He-Ne and Argon lasers.

- Lasers have wide range of industrial applications for cutting, and boring large metal objects, and applications in medicine, scientific research, technological devices, communication and holography.

- Holography is the production of a 3-D image, known as hologram, of an object using lasers which exploits the phase difference of the reflected rays from the object. A hologram is like a photographic negative of the image, the image not being visible to bare eye. Lasers from the same source are then used to project the 3-D image of the object from the hologram into the space.

9.8 Exercises

9.1. Using Bohr's theory of the hydrogen atom, calculate the following:
(i) the radius of an electron in the second Bohr orbit?

(ii) the velocity of the electron in the second orbit.

(iii) the frequency of the photon emitted when an electron makes a transition from the energy level $n = 2$ to the energy level $n = 1$, assuming that the energy for the nth level is given by

$$E_n = -\frac{13.6}{n^2} \text{ eV}$$

9.2. Calculate the wavelengths and frequencies of the least energetic and the most energetic spectral lines in the Lyman. Balmer, Paschen, Bracket and Pfund series of the hydrogen spectrum. In what regions of electromagnetic spectrum do these lines lie?

9.3. Which spectral series of the hydrogen spectrum have spectral lines in the visible region? Calculate the wavelength and frequencies of the first and the last hydrogen spectral lines in the visible region in the spectrum of hydrogen. What are the energy states (n_i, n_f) that produce these spectral lines, and identify the corresponding spectral series?

9.4. Calculate the frequency and photon energy of all the laser lines given in Table 9.1. What is the energy difference between the levels that produces these laser lights?

9.5. In your library search the literature on the applications of lasers and write a two page (1000 words) essay on at least 10 different applications of lasers. Name at least five applications (different from the essay) of lasers which one comes across almost on regular basis in every day life.

9.6. Where lasers are in use, there is usually a danger sign cautioning people not to look into the beam and/ or not to cross the path of a laser beam. Do you know what are the dangers of laser light, and why and in what way they are harmful? Search literature in your library to find the answer and write one page essay on it.

9.7. A 100 watts laser emits at 10.2 μm wavelength radiation. Calculate the rate at which the photons are produced by the laser. If the laser beam is 0.5 mm wide, calculate the power per unit area delivered by the laser.

9.8. In what respect a hologram negative and the image produced from it differ from the negative and the image produced by ordinary photograph.

9.9. Calculate the radius of the first Bohr orbit of the hydrogen atom, and the velocity of the electron in it. Calculate the corresponding de Broglie wavelength of the electron, and its ratio to the circumference of the orbit.

9.10. Find a general expression relating the de Broglie wavelength to the circumference of the quantized electronic orbits in the atom. From your results, formulate a quantization rule in terms of the de Broglie wavelength of the electrons in atomic orbits.

9.11 Verify that the quanties $\eta(\omega)$, $\kappa(\omega)$, $\epsilon'(\omega)$, $\epsilon''(\omega)$ defined in section 9.32 satisfy the following relations given in equations (9.35) and (9.36).

$$\eta(\omega) = \{\frac{1}{2}[\epsilon'(\omega) + \sqrt{(\epsilon'(\omega)^2 + \epsilon''(\omega)^2)}]\}^{1/2}$$

$$\kappa(\omega) = \{\frac{1}{2}[-\epsilon'(\omega) + \sqrt{(\epsilon'(\omega)^2 + \epsilon''(\omega)^2)}]\}^{1/2}$$

9.12. (a) Show that the Bohr radius $(= 4\pi\epsilon_0\hbar^2/m_e e^2)$, the Compton wavelength $(h/m_e c)$ divided by 2π, and the classical electron radius $(= e^2/4\pi\epsilon_0 m_e c^2)$ are in the ratio $1 : \alpha : \alpha^2$, where $\alpha = e^2/4\pi\epsilon_0\hbar c$. The quantity α is known as the *fine-structure constant*.
(b) Calculate the value of α and $1/\alpha$.

9.13. The characteristic red light of a He-Ne laser is due to stimulated emission between neon levels at 20.66 eV and 18.70 eV. Calculate the wavelength and frequency of the emitted radiation.

9.14. A laser emits radiation at a wavelength of 555 nm. Photons are emitted at the rate of 8.5×10^{18} s^{-1}. What is the power of the laser?

9.15. The tube of a laser has a mirror at each end. A standing electromagnetic wave fills the space between these two mirrors. This wave must satisfy the boundary condition that its electric field is zero at the position of the mirrors.
(a) Show that the distance L between the mirrors and the wavelength λ of the wave are related by

$$L = \left(\frac{n+1}{2}\right)\lambda \qquad\qquad n = 0,1,2,\cdots$$

(b) Calculate the value of n for a He-Ne laser with $L = 30.0$ cm and $\lambda = 6328\mathring{A}$.

Appendix A

Some Useful Constants etc

Table A.1: Some Fundamental Physical Constants

Name	Symbol	Value	Units
Acceleration due to gravity	g	$9.807^{\dagger}$	ms^{-2}
Universal gravitational constant	G	6.673×10^{-11}	Nm^2kg^{-2}
Electronic charge	e	1.602×10^{-19}	C
Rest mass of electron	m_e	9.109×10^{-31}	kg
Rest mass of proton	m_p	1.673×10^{-27}	kg
Rest mass of neutron	m_n	1.675×10^{-27}	kg
Mass of hydrogen atom	m_H	1.673×10^{-27}	kg
Bohr radius	a_o	0.529×10^{-10}	m
Binding energy of hydrogen	E_o	13.606	eV
Speed of light in vacuum	c	2.998×10^8	ms^{-1}
Permittivity of free space	$\epsilon_o = \frac{1}{\mu_o c^2}$	8.854×10^{-12}	$C^2(Nm^2)^{-1}$
Permeability of free space	μ_o	$4\pi \times 10^{-7}$	NA^{-2}
Coulomb's constant for free space	$k = \frac{1}{4\pi\epsilon_o}$	8.988×10^9	Nm^2C^{-2}
Planck's constant	h	6.626×10^{-34}	Js
	$\hbar = \frac{h}{2\pi}$	1.055×10^{-34}	Js
Stefan-Boltzmann constant	σ	5.67×10^{-8}	$Wm^{-2}K^{-4}$
Boltzmann's constant	$k = \frac{R}{N_A}$	1.381×10^{-23}	JK^{-1}
Avogadro's number	N_A	6.022×10^{26}	$(kmol)^{-1}$
Universal gas constant	R	8.315	$J(mol-K)^{-1}$
Standard temperature	T	273.15	K
Standard pressure	P	1	$atmosphere$

$\dagger$: Average value at sea level; decreases to $9.598\ ms^{-2}$ at an altitude of $100,000\ m$. Varies with latitude from $9.780\ ms^{-2}$ at the equator to $9.832\ ms^{-2}$ at the poles.

Table A.2: Some common physical properties and parameters

Name	Symbol	Value	Units
Volume of ideal gas at STP	V	22.415×10^3	$l \, (kmol)^{-1}$
Density of air at STP	ρ_{air}	1.293	kgm^{-3}
Average molecular weight of air	m_{air}	28.970	$kg(kmol)^{-1}$
Density of water at 4^oC	ρ_w	1000	kgm^{-3}
Specific heat capacity of water	C_w	4200	Jkg^{-1}
Latent heat of fusion of ice	L_f	3.360×10^5	Jkg^{-1}
Latent heat of vaporization of water	L_v	2.260×10^6	Jkg^{-1}
Speed of sound in air at STP	$v_{s,air}$	331	ms^{-1}
Absolute zero (0 K) temperature	K	-273.15	^oC

Table A.3: Earth, Sun and Moon data

Earth	Mass	$5.98 \times 10^{24} kg$
	Mean radius	$6.378 \times 10^6 m$
	Period of revolution (sun)	$1y = 365d\ 5h\ 48m\ 45.97s$
	Period of rotation (axial)	$86164.1s = 23h\ 56m\ 4.1s$
	Mean Earth - sun distance	$1.496 \times 10^{11} m$
	Mean Earth - moon distance	$3.844 \times 10^8 m$
Moon	Mass	$7.35 \times 10^{22} kg$
	Mean radius	$1.738 \times 10^6 m$
	Period of revolution (earth)	$27.32\ days$
Sun	Mass	$1.99 \times 10^{30} kg$
	Mean radius	$6.96 \times 10^8 m$
	Period of rotation (axial)	$25 - 35\ days$

Table A.4: Some conversion factors

Fundamental units:		
	1 $slug$ =	14.59 kg
	1 $slug$ =	32.17 lb mass where $g = 9.81\ ms^{-2}$
	1 kg =	2.20 lb mass where $g = 9.81\ ms^{-2}$
	1 $year\ (y)$ =	365.242(19878) $days = 3.156 \times 10^7 s$
	1 in =	2.54 cm
	1 mi =	5280 $ft = 1.61\ km$
	1 $nautical\ mile\ (U.S.)$ =	1.151 $mi = 6074\ ft = 1.852\ km$

Secondary units:

Speed	1 $knot$ =	1.852 $kmh^{-1} = 1.151\ mih^{-1}$
Volume	1 $gallon\ (= 4qt)\ (U.S.)$ =	3.78 $l = 0.83\ gal\ (Imperial)$

Atomic mass unit (u)	1 u =	$1.661 \times 10^{-27} kg = 931.434\ MeV\ c^{-2}$
Electron volt (eV)	1 eV =	$1.602 \times 10^{-19} J$
Mechanical equivalent of heat (J)	J =	$4.185\ J\ (cal)^{-1} = 3.97 \times 10^{-3}\ Btu\ (cal)^{-1}$
Angstrom $(\mathring{A})$	1 $\mathring{A}$ =	$1 \times 10^{-10} m$
Fermi (fm)	1 fm =	$1 \times 10^{-15} m$
Light-year (ly)	1 ly =	$9.462 \times 10^{15} m$
Atmospheric pressure (atm)	1 atm =	$1.013 \times 10^5 Nm^{-2} = 1.013\ bar = 760\ torr$
	1 $radian\ (rad)$ =	$57.30^o = 57^o\ 18'$

Table A.5: The Greek Alphabet

Alpha	A	α		Nu	N	ν
Beta	B	β		Xi	Ξ	ξ
Gamma	Γ	γ		Omicorn	O	o
Delta	Δ	δ		Pi	Π	$\pi,\ \varpi$
Epsilon	E	$\epsilon,\ \varepsilon$		Rho	P	$\rho,\ \varrho$
Zeta	Z	ζ		Sigma	Σ	$\sigma,\ \varsigma$
Eta	H	η		Tau	T	τ
Theta	Θ	$\theta,\ \vartheta$		Upsilon	Υ	υ
Iota	I	ι		Phi	Φ	$\Phi,\ \varphi$
Kappa	K	κ		Chi	X	χ
Lambda	Λ	λ		Psi	Ψ	ψ
Mu	M	μ		Omega	Ω	ω

Appendix B

Mathematical Relations

B.1. Mathematical Symbols and Constants

Table B.1.1. Mathematical symbols and numerical constants

Expression	Explanation
$a \propto b$	a is prpopotional to b
$a \approx b$	a is approximately equal to b
$a \sim b$	a is same order of magnitude (*i.e.* same power of 10) as b
$a \neq b$	a is not equal to b
$a \perp b$	a is perpendicular to b
$a \parallel b$	a is parallel to b
$a > b$	a is greater than b
$a < b$	a is less than b
$a \geq b$	a is greater than or equal to b OR a is not less than b
$a \leq b$	a is less than or equal to b OR a is not greater than b
$a \gg b$	a is much larger than b
$a \ll b$	a is much less (smaller) than b
$\sum_i a_i$	Summation: $a_1 + a_2 + a_3 + ...$
$\prod_i a_i$	Product: $a_1 \times a_2 \times a_3 \times ...$
n!	Factorial n(integer): $n \times (n-1) \times (n-2)... \times 3 \times 2 \times 1.$
Δx	Small change in x
$\Delta x \to 0$	Δx approaches zero
$\bar{x}$	Average value of x
∞	Infinity

Constant	Value
π(rad)	$3.14159\ 26535\ 89793...$ rad $= 180^o$
1 rad	$57.2957795^o = 57^o17'44.8''$
e	$2.7182\ 818...$
ln 2	$0.6931472...$
ln 10	$2.3025851...$
$\log_{10} e$	$0.4342945...$

B.2. Algebra

NOTE: In this section some of the common algebraic mistakes made by students are also included as a cautionary measure. These are expressed using the symbol $\neq$.

Quadratic equation

The quadratic equation of the form: $ax^2 + bx + c = 0$ has two roots given by:

$$x = \frac{(-)b \pm \sqrt{b^2 - 4ac}}{2a}$$

Ratios, fractions, and inversion

$$\text{If } \frac{a}{b} = \frac{c}{d}, \text{ then}$$

$$\frac{a}{c} = \frac{b}{d},$$

$$\frac{b}{a} = \frac{d}{c}, \text{ and}$$

$$\frac{c}{a} = \frac{d}{b}$$

$$\text{If } a = (b + c + d), \text{ then}$$

$$\frac{1}{a} = \frac{1}{(b + c + d)}, \text{ and}$$

$$\frac{1}{a} \neq \frac{1}{a} + \frac{1}{b} + \frac{1}{c}$$

$$\text{If } \frac{a}{b} = \frac{c}{d} + \frac{e}{f}, \text{ then}$$

$$\frac{b}{a} = \frac{1}{\{\frac{c}{d} + \frac{e}{f}\}}, \text{ and}$$

$$\frac{b}{a} \neq \frac{d}{c} + \frac{f}{e}$$

Algebraic expansions and factorization

$$(x \pm a)^2 = x^2 \pm 2ax + a^2$$
$$(x \pm a)^3 = x^3 \pm 3ax^2 + 3a^2x \pm a^3$$
$$(x \pm a)^n = x^n + C_1an^{n-1} + C_2a^2x^{n-2} + ... + C_ja^jx^{(n-j)} + ...C_na^n$$

$$\text{where} C_j = (\pm 1)^j \frac{\{n(n-1)(n-2)...(n-j+1)\}}{j!}$$

$$
\begin{aligned}
(x^2 - a^2) &= (x + a)(x - a) \\
(x^3 \pm a^3) &= (x \pm a)(x^2 \mp ax + a^2) \\
(x^4 - a^4) &= (x^2 + a^2)(x^2 - a^2) = (x^2 + a^2)(x + a)(x - a)
\end{aligned}
$$

Manupulation of the powers and roots

$$
\begin{aligned}
x^0 &= 1 \\
x^1 &= x \\
\frac{1}{x^n} &= x^{-n} \\
x^{\frac{1}{n}} &= \sqrt[n]{x} \\
x^n . x^m &= x^{n+m} \\
(x^n)^m &= x^{nm} \\
\sqrt{(a^2 + b^2 + c^2)} &\neq (a + b + c)
\end{aligned}
$$

Logrithmic and exponential functions

$$
\begin{aligned}
\ln x &= \ln_e x \\
\text{If } \ln x &= y, \text{ then} \\
x &= e^y \\
\ln e &= 1 \\
\ln 1 &= 0 \\
\ln 10 &= 2.302585...
\end{aligned}
$$

$$
\begin{aligned}
\ln (ab) &= \ln a + \ln b \\
\ln \frac{a}{b} &= \ln a - \ln b \\
\ln a^b &= b \ln a \\
\text{If } z &= ae^{bx}, \text{ then} \\
\ln z &= b\,x + \ln a
\end{aligned}
$$

$$
\begin{aligned}
\log x &= \log_{10} x \\
\text{If } \log x &= y, \text{ then} \\
y &= 10^x \\
\log 1 &= 0 \\
\log 10 &= 1 \\
\log 100 &= 2
\end{aligned}
$$

$$\log 10^n = n$$
$$\ln f(x) = \ln 10 \times \log f(x)$$
$$= (2.302585) \times \log f(x)$$

$$e^{\pm ax} = 1 \pm ax + \frac{(ax)^2}{2!} \pm \frac{(ax)^3}{3!} + ...(\pm 1)^j \frac{(ax)^j}{j!}..., \qquad (j = 0, 1, 2, 3, ...\infty)$$

For $(-)1 < x < 1$,

$$\ln(1+x) = x - \frac{1}{2}x^2 + \frac{1}{3}x^3 - \frac{1}{4}x^4 + \frac{1}{5}x^5 - ...$$
$$\ln(1-x) = -x - \frac{1}{2}x^2 - \frac{1}{3}x^3 - \frac{1}{4}x^4 - \frac{1}{5}x^5 - ...$$

Approximations

For $x \ll 1$:

$$e^{\pm x} \cong (1 \pm x)$$
$$\ln(1 \pm x) \cong \pm x$$
$$(1 \pm x)^n \cong (1 \pm nx)$$
$$(1 \pm x)^{\frac{1}{2}} \cong 1 \pm \frac{1}{2}x$$
$$(1 \pm x)^{-1} \cong 1 \mp x$$
$$(1 \pm x)^{-\frac{1}{2}} \cong 1 \mp \frac{1}{2}x$$

B.3. Plane Geometry

Triangles

For a general triangle of sides *a*, *b*, and *c* and opposite angles *A*, *B*, and *C*:

$$A + B + C = 180° = \pi \text{ radians}$$
$$c^2 = a^2 + b^2 - 2\,a\,b\cos C$$

$$\text{Area of the triangle} = \frac{1}{2}a\,b\sin C = \frac{1}{2}b\,c\sin A = \frac{1}{2}c\,a\sin B$$
$$= \sqrt{s \times (s-a) \times (s-b) \times (s-c)}$$
$$\text{where } s = \frac{1}{2}(a+b+c)$$

$$\frac{a}{\sin A} = \frac{b}{\sin B} = \frac{c}{\sin C}$$

For a right angled trangle in which $\angle C = 90°$, the opposite side c is called the hypotenous, and $c^2 = a^2 + b^2$. This relationship between the sides of a right angle triangle is called the *Pythagora's Theorem*.

Straight line:

General equation of a straight line is: $y = mx + c$.
where c is the $y-$ intercept of the line, *i.e.* the value of y when $x = 0$, and
m is the slope of the line given as: $m = \frac{y_2 - y_1}{x_2 - x_1} = \Delta y / \Delta x = \tan \theta$. The angle θ is measured from the positive $x-$ axis and is positive anticlockwise and negative clockwise.

Angle between two straight lines of slope m_1 and m_2 is given by:

$$\tan \alpha = \frac{m_2 - m_1}{1 + m_1 m_2}$$
$$\text{For parallel lines: } m_1 = m_2, \text{ and}$$
$$\text{for perpendicular lines: } m_1 \times m_2 = (-)1.$$

Equation of a straight line passing through points (x_1, y_1) and (x_2, y_2) is:
$y = m(x - x_1) + y_1 = m(x - x_2) + y_2$ where $m = \frac{(y_2 - y_1)}{(x_2 - x_1)}$

Distance d between two points (x_1, y_1) and (x_2, y_2) is given by: $[(x_2 - x_1)^2 + (y_2 - y_1)^2]^{\frac{1}{2}}$

Circle

The general equation of a circle is: $x^2 + y^2 + 2bx + 2cy + d = 0$
with center at $(-b, -c)$ and radius, $r = \sqrt{b^2 + c^2 - d}$

Equation of a circle with center at (x_c, y_c) and radius R is: $(x - x_c)^2 + (y - y_c)^2 = R^2$

Area of a circle of radius R is: πR^2, and its circumfrence is: $2 \pi R$ where $\pi = 3.1415...$ radians.

Area of a sector of angle θ (radians) of a circle of radius R is: $\frac{1}{2}\theta R^2$ and the arc of the sector is: θR

Ellipse

The general equation of an ellipse is: $Ax^2 + Bxy + Cy^2 + Dx + Ey + F = 0$ where $(B^2 - 4AC) < 0$.

Equation of an ellipse with center at the origin, and the axes of the ellipse of lengths $2a$ and $2b, (a \neq b)$, along the $x-$ and $y-$ axes respectively is: $\frac{x^2}{a^2} + \frac{y^2}{b^2} = 1$.
Area of the ellipse is: $\pi a b$ and its circumfrence is $\approx 2\pi \left[\frac{1}{2}(a^2 + b^2)\right]^{\frac{1}{2}}$

Equation of an ellipse with center at (x_o, y_o) and axes of the ellipse parallel to the coordinate axes is: $\frac{(x - x_o)^2}{a^2} + \frac{(y - y_o)^2}{b^2} = 1$

Longer of the two axis is know as the major axis, and the shorter is the minor axis. Let $2a$ be the major axis and $2b$ be the minor axis, $\{(a/b) > 1\}$, then a and b are the semi-major and semi-minor axes respectively.

The eccentricity (e) of the ellipse is given by: $e = \frac{\sqrt{(a^2 - b^2)}}{a} < 1$

Circle is a special case of an ellipse for which $a = b, (a/b) = 1$, ecentricity $e = 0$, and the radius of the circle is $R = a = b$

Parabola

Eccentricity of a parabola: $e = 1$.

The general equation of a parabola is: $Ax^2 + Bxy + Cy^2 + Dx + Ey + F = 0$ where $(B^2 - 4AC) = 0$, and the axis of the parabola is oblique to the coordinate axes.

The equation of a parabola with its axis parallel to the $x-$ axis is: $x = ay^2 + by + c$.

The equation of a parabola with its axis parallel to the $y-$ axis is: $y = ax^2 + bx + c$.

The equation of a parabola with its axis along the $x-$ axis, vertex at the origin, and focus at $(p, 0)$ is: $y^2 = 4px$.

Hyperbola

Let $2a$ be the transverse axis, and $2b$ be the conjugate axis of the hyperbola. The eccentricity of a hyperbola is given by: $e = \frac{\sqrt{(a^2 + b^2)}}{a} > 1$

For a rectangular hyperbola $a = b$ and $e = \sqrt{2}$, and the asymptotes are perpendicular.

Genaral equation of a hyperbola with axes oblique to the coordinate axes is: $Ax^2 + Bxy + Cy^2 + Dx + Ey + F = 0$ where $(B^2 - 4AC) > 1$

Equation of a hyperbola with center at the origin, and the transverse axis of length $2a$ and and the conjugate axis of length $2b$ along the $x-$ and $y-$ axes respectively is: $\frac{x^2}{a^2} - \frac{y^2}{b^2} = 1$.

Equation of an ellipse with center at (x_o, y_o), and the transverse axis $2a$ and the conjugate axis $2b$ parallel to the $x-$ and $y-$ axes respectively is: $\frac{(x-x_o)^2}{a^2} - \frac{(y-y_o)^2}{b^2} = 1$.

Transformation of coordinates

Let the coordinates of a point P in a 3-D space in cartesian, spherical and cylinderical coordinates system be: $P = (x, y, z) = (r, \theta, \phi) = (\rho, \phi, z)$ respectively. Then the spherical and the cylindrical coordinates are related to the cartesian coordinates by the following expressions.

Spherical and cartesian coordinates:

$$
\begin{aligned}
x &= r \sin\theta \cos\phi \\
y &= r \sin\theta \sin\phi \\
z &= r \cos\theta
\end{aligned}
$$

$$
\begin{aligned}
r &= \sqrt{x^2 + y^2 + z^2} \\
\theta &= \cos^{-1}\left(\frac{z}{\sqrt{x^2 + y^2 + z^2}}\right) = \cos^{-1}\left(\frac{z}{r}\right) \\
\phi &= \tan^{-1}\left(\frac{y}{x}\right)
\end{aligned}
$$

Cylindrical and cartesian coordinates:

$$x = \rho \cos \phi$$
$$y = \rho \sin \phi$$
$$z = z$$

$$\rho = \sqrt{x^2 + y^2}$$
$$\phi = \tan^{-1}\left(\frac{y}{x}\right)$$
$$z = z$$

Volume element in the three coordinate systems:
Cartesian coordinates: $dv = dx\, dy\, dz$
Spherical coordinates: $dv = r^2\, dr \sin \theta\, d\theta\, d\phi$
Cylindrical coordinates: $dv = \rho d\rho d\phi dz$

Surface element in the three coordinate systems:
Cartesian coordinates: $ds = dx\, dy$
Spherical coordinates: $ds = R^2 \sin \theta\, d\theta\, d\phi$
Cylindrical coordinates: $ds = R\, d\phi\, dz$
where R is the radius of the spherical (cylinderical) surface.

B.4. Solid (3-D) geometry

The following symbols are used for expressions in this section:
V = volume,
T = total surface area,
S = lateral surface area, and
s = surface area of one of the face where all faces are equal.

Parallelepiped

Let the three edges of the parallelepiped are given by vectors **a, b,** and **c**, and the dihedral angles are $\alpha \neq \beta \neq \gamma$.

$V = \mathbf{a}.(\mathbf{b} \times \mathbf{c}) = \mathbf{b}.(\mathbf{c} \times \mathbf{a}) = \mathbf{c}.(\mathbf{a} \times \mathbf{b})$
$T = 2 \times |(\mathbf{a} \times \mathbf{b}) + (\mathbf{b} \times \mathbf{c}) + (\mathbf{c} \times \mathbf{a})| = 2(a\, b\, \sin \gamma + b\, c\, \sin \alpha + c\, a\, \sin \beta)$

For a rectangular parallelepiped:
$\alpha = \beta = \gamma = 90^o$,
$V = a \times b \times c$,
$T = 2(ab + bc + ca)$, and
body diagonal: $D = (a^2 + b^2 + c^2)^{\frac{1}{2}}$.

Tetrahedron

A solid bounded by four equalateral triangles of side a each.

Height: $h = a\sqrt{\frac{2}{3}}$, $s = \frac{\sqrt{3}}{4}a^2$, and $V = \frac{1}{6\sqrt{2}}a^3$.

Pyramid

$V = \frac{1}{3}$(area of base) × (altitude)

$T =$ (area of slant surfaces) + (area of base) $= \frac{1}{2}$(perimeter of base) × (slant height) + (area of base).

Cone

Cone is a regular pyramid with a circular base of radius R, and height h.

$V = \frac{1}{3}\pi R^2 h$

Stant height $h_s = (R^2 + h^2)^{\frac{1}{2}}$

$S = \pi R h_s = \pi R(R^2 + h^2)^{\frac{1}{2}}$.

$T = S + \pi R^2$.

Cylinder

For a right circular cylinder† of base radius R, and height h.

$V = \pi R^2 h$, cylinder surface area: $S = 2\pi R h$, and $T = \pi R \times (R + 2h)$

$\dagger$ Same relations apply to a disk of radius R and thickness t whereby h is replaced with t.

Sphere

$V = \frac{4}{3}\pi R^3$, and $T = 4\pi R^2$, where R is the radius of the sphere.

Ellipsoid

$V = \frac{4}{3}\pi\, a\, b\, c$, where a, b, and c are the semi axes of the ellipsoid.

Spheroid

Formed by rotating an ellipse of major and minor semiaxes a and b respectively and ecentricity e about one of the axes.

Oblate spheroid: Rotation about the minor axis: $V = \frac{4}{3}\pi a^2 b$, and $T = 2\pi a^2 + \pi\, \frac{b^2}{e}\, \ln\frac{(1+e)}{(1-e)}$

Prolate spheroid: Rotation about the major axis: $V = \frac{4}{3}\pi a b^2$, and $T = 2\pi b^2 + 2\pi\, \frac{ab}{e}\, \sin^{-1} e$

B.5. Trigonometry

Consider a right angle triangle ABC with sides a, b and c. $\angle C = 90^o$, $\angle A = \theta$ and $\angle B = (90 - \theta)^o$. The side c of the triangle is the hypotenuse, and with respect to $\angle A = \theta$, a is designated as the opposite side, and b is designated as the adjacent side. The three sides of the triangle are related by the Pythagora's theorem: $c^2 = (a^2 + b^2)$. Using this nomenclature, we define the following trigonometric functions for $\angle A = \theta$.

$$\text{sine } \theta \;\; = \;\; \sin\theta = \frac{a}{c} = \frac{\text{opposite}}{\text{hypotenuse}}$$

$$\text{cosine } \theta \;\; = \;\; \cos \theta = \frac{b}{c} = \frac{\text{adjacent}}{\text{hypotenuse}}$$

$$\text{tangent } \theta \;\; = \;\; \tan \theta = \frac{a}{b} = \frac{\text{opposite}}{\text{adjacent}}$$

and

$$\text{cosecant } \theta \;\; = \;\; \text{cosec } \theta = \frac{1}{\sin \theta}$$

$$\text{secent } \theta \;\; = \;\; \sec \theta = \frac{1}{\cos \theta}$$

$$\text{cotangent } \theta \;\; = \;\; \cot \theta = \frac{1}{\tan \theta}$$

Table of trigonometric functions

θ^o	$\sin \theta$	$\cos \theta$	$\tan \theta$
0	0	1	0
30	$\frac{1}{2}$	$\frac{\sqrt{3}}{2}$	$\frac{1}{\sqrt{3}}$
45	$\frac{1}{\sqrt{2}}$	$\frac{1}{\sqrt{2}}$	1
60	$\frac{\sqrt{3}}{2}$	$\frac{1}{2}$	$\sqrt{3}$
90	1	0	∞
In the I^{st} quadrant	(+)	(+)	(+)
In the IInd quadrant	(+)	(-)	(-)
In the IIIrd quadrant	(-)	(-)	(+)
In the IVth quadrant	(-)	(+)	(-)
$(-)\alpha$	(-) $\sin \alpha$	(+) $\cos \alpha$	(-) $\tan \alpha$
$90 \pm \alpha$	(+) $\cos \alpha$	($\mp$) $\sin \alpha$	($\mp$) $\cot \alpha$
$180 \pm \alpha$	($\mp$) $\sin \alpha$	(-) $\cos \alpha$	($\pm$) $\tan \alpha$
$270 \pm \alpha$	(-) $\cos \alpha$	($\pm$) $\sin \alpha$	($\mp$) $\cot \alpha$
$360 \pm \alpha$	($\pm$) $\sin \alpha$	(+) $\cos \alpha$	($\pm$) $\tan \alpha$

Trigonometric identities

$$\sin^2 \theta + \cos^2 \theta \;\; = \;\; 1$$

$$\sec^2 \theta \;\; = \;\; 1 + \tan^2 \theta$$

$$\text{cosec}^2 \theta \;\; = \;\; 1 + \cot^2 \theta$$

$$\sin (\alpha \pm \beta) \;\; = \;\; \sin \alpha \cos \beta \pm \cos \alpha \sin \beta$$

$$\cos (\alpha \pm \beta) \;\; = \;\; \cos \alpha \cos \beta \mp \sin \alpha \sin \beta$$

$$\tan (\alpha \pm \beta) \;\; = \;\; \frac{(\tan \alpha \pm \tan \beta)}{(1 \mp \tan \alpha \tan \beta)}$$

$$\sin 2\alpha = 2\sin\alpha\cos\alpha$$

$$\cos 2\alpha = \cos^2\alpha - \sin^2\alpha = 2\cos^2\alpha - 1 = 1 - 2\sin^2\alpha$$

$$\tan 2\alpha = \frac{2\tan\alpha}{(1 - \tan^2\alpha)}$$

$$\sin 3\alpha = 3\sin\alpha - 4\sin^2\alpha$$

$$\cos 3\alpha = 4\cos^2\alpha - 3\cos\alpha$$

$$\tan 3\alpha = \frac{3\tan\alpha - \tan^3\alpha}{(1 - 3\tan^2\alpha)}$$

$$\sin\alpha + \sin\beta = 2\sin\left(\frac{\alpha+\beta}{2}\right)\cos\left(\frac{\alpha-\beta}{2}\right)$$

$$\sin\alpha - \sin\beta = 2\cos\left(\frac{\alpha+\beta}{2}\right)\sin\left(\frac{\alpha-\beta}{2}\right)$$

$$\cos\alpha + \cos\beta = 2\cos\left(\frac{\alpha+\beta}{2}\right)\cos\left(\frac{\alpha-\beta}{2}\right)$$

$$\cos\alpha - \cos\beta = (-)2\sin\left(\frac{\alpha+\beta}{2}\right)\sin\left(\frac{\alpha-\beta}{2}\right)$$

$$\tan\alpha + \tan\beta = \frac{\sin(\alpha+\beta)}{\cos\alpha\cos\beta}$$

$$\tan\alpha - \tan\beta = \frac{\sin(\alpha-\beta)}{\cos\alpha\cos\beta}$$

$$\sin\alpha\sin\beta = \frac{1}{2}\cos(\alpha-\beta) - \frac{1}{2}\cos(\alpha+\beta)$$

$$\cos\alpha\cos\beta = \frac{1}{2}\cos(\alpha-\beta) + \frac{1}{2}\cos(\alpha+\beta)$$

$$\sin\alpha\cos\beta = \frac{1}{2}\sin(\alpha+\beta) + \frac{1}{2}\sin(\alpha-\beta)$$

$$\cos\alpha\sin\beta = \frac{1}{2}\sin(\alpha+\beta) - \frac{1}{2}\sin(\alpha-\beta)$$

For angle α in radians:

$$\sin\alpha = \frac{e^{i\alpha} - e^{-i\alpha}}{2i}; \quad \text{where } i = \sqrt{-1}$$

$$\cos\alpha = \frac{e^{i\alpha} + e^{-i\alpha}}{2}$$

$$e^{\pm i\alpha} = \cos\alpha \pm i\sin\alpha$$

$$\sin\alpha = \alpha - \frac{\alpha^3}{3!} + \frac{\alpha^5}{5!} - \frac{\alpha^7}{7!} + ... + (-1)^{2j+1}\frac{\alpha^{2j+1}}{(2j+1)!} + ..., \quad \text{where } j = 0, 1, 2, 3, ...$$

$$\cos\alpha = 1 - \frac{\alpha^2}{2!} + \frac{\alpha^4}{4!} - \frac{\alpha^6}{6!} + ... + (-1)^{j}\frac{\alpha^{2j}}{(2j)} + ...$$

For α (radians) $\ll 1$: $\sin \alpha = \alpha$; $\cos \alpha = 1$; $\tan \alpha = \alpha$

Hyperbolic functions

For a real argument u:

$$\text{hyperbolic sine of u} \quad = \quad \sinh u = \frac{e^u - e^{-u}}{2}$$

$$\text{hyperbolic cosine of u} \quad = \quad \cosh u = \frac{e^u + e^{-u}}{2}$$

$$\text{hyperbolic tangent of u} \quad = \quad \tanh u = \frac{\sinh u}{\cosh u}$$

$$\text{hyperbolic cosecant of u} \quad = \quad \text{cosech } u = \frac{1}{\sinh u}$$

$$\text{hyperbolic secant of u} \quad = \quad \text{sech } u = \frac{1}{\cosh u}$$

$$\text{hyperbolic cotangent of u} \quad = \quad \coth u = \frac{1}{\tanh u}$$

Relationship between the squares of functions

$$\cosh^2 u - \sinh^2 u \quad = \quad 1$$
$$\tanh^2 u + \text{sech}^2 u \quad = \quad 1$$
$$\coth^2 u - \text{cosech}^2 u \quad = \quad 1$$
$$\text{cosech}^2 u - \text{sech}^2 u \quad = \quad \text{cosech}^2 u \,\text{sech}^2 u$$

Symmetry and periodicity

$$\sinh (-u) \quad = \quad (-)\sinh u$$
$$\cosh (-u) \quad = \quad \cosh u$$
$$\tanh (-u) \quad = \quad (-)\tanh u$$
$$\text{cosech} (-u) \quad = \quad (-)\text{cosech } u$$
$$\text{sech} (-u) \quad = \quad \text{sech } u$$
$$\coth (-u) \quad = \quad (-)\coth u$$

Range of functions for real argument u

Function	Range of u	Range of function
sinh u	$(-\infty, +\infty)$	$(-\infty, +\infty)$
cosh u	$(-\infty, +\infty)$	$(1, +\infty)$
tanh u	$(-\infty, +\infty)$	$(-1, +1)$
cosech u	$(-\infty, 0)$	$(0, -\infty)$
	$(0, +\infty)$	$(+\infty, 0)$
sech u	$(-\infty, +\infty)$	$(0, 1)$
coth u	$(-\infty, 0)$	$(1, -\infty)$
	$(0, +\infty)$	$(+\infty, 1)$

Special values of hyperbolic functions

x	0	$\frac{\pi}{2}i$	πi	$\frac{3\pi}{2}i$	∞
sinh x	0	i	0	$-i$	∞
cosh x	1	0	-1	0	∞
tanh x	0	∞i	0	$-\infty i$	1
cosech x	∞	$-i$	∞	i	0
sech x	1	∞	-1	∞	0
coth x	∞	0	∞	0	1

B.6. Differential Calculus

NOTE: In the differential expressions given below, and in the integral expressions in the next section A, B and C are the functions of x, $f(A)$ is a function of A, and a, b, c, n, and m are real constants. The arguments of trigonometric functions are expressed in radians.

Definition:

$$\frac{d}{dx}A = Lim_{\Delta x \to 0} \frac{\Delta A}{\Delta x}$$

$$\frac{d}{dx}(A \pm B \mp C) = \frac{d}{dx}A \pm \frac{d}{dx}B \mp \frac{d}{dx}C$$

$$\frac{d}{dx}(ABC) = BC\frac{d}{dx}A + CA\frac{d}{dx}B + AB\frac{d}{dx}C$$

$$\frac{d}{dx}\left(\frac{1}{ABC}\right) = \frac{1}{BC}\frac{d}{dx}\frac{1}{A} + \frac{1}{CA}\frac{d}{dx}\frac{1}{B} + \frac{1}{AB}\frac{d}{dx}\frac{1}{C}$$

$$\frac{d}{dx}\left(\frac{A}{BC}\right) = \frac{1}{BC}\frac{d}{dx}A + A\frac{d}{dx}\frac{1}{BC}$$

$$\frac{d}{dx}f(A) = \frac{d}{dA}f(A)\frac{d}{dx}A$$

$$\frac{d}{dx}A^B = B\,A^{B-1}\frac{d}{dx}A + A^B \ln A\frac{d}{dx}B$$

Table of some common differential expressions

A(x)	$\frac{d}{dx}$ A
$a\,x^n$	$a\,nx^{n-1}$
a^x	$a^x \ln a$
e^x	e^x
$\ln x$	$\frac{1}{x}$
$\sin x$	$\cos x$
$\cos x$	$(-)\sin x$
$\tan x$	$\sec^2 x$
$\mathrm{cosec}\, x$	$(-)\cot x\,\mathrm{cosec}\, x$
$\sec x$	$\tan x\,\sec x$
$\cot x$	$(-)\,\mathrm{cosec}^2 x$

f(A)	$\frac{d}{dx}$ f(A)
$a\,A^n$	$a\,nA^{(n-1)}\,\frac{d}{dx}A$
a^A	$a^A \ln a\,\frac{d}{dx}A$
e^A	$e^A\,\frac{d}{dx}A$
$\ln A$	$\frac{1}{A}\,\frac{d}{dx}A$
$\sin A$	$\cos A\,\frac{d}{dx}A$
$\cos A$	$(-)\sin A\,\frac{d}{dx}A$
$\tan A$	$\sec^2 A\,\frac{d}{dx}A$

$$d(ABC) = BC\,dA + CA\,dB + AB\,dC$$

If $y = f(x)$, then $dy = \frac{d}{dx}f(x)\,dx$

B.7. Integral Calculus

Definition:

$$\sum_{Lim\,\Delta x_i \to 0}(A(x_i)\,\Delta x_i) = \int A(x)\,dx$$

Integral of the product of two functions AB is obtained by **integration by parts** following the expression given below:

$$\int (AB)\,dx = A\int B dx - \int\left(\frac{d}{dx}A\int B\,dx\right)dx$$

Table of some common integral expressions

A(x)	$\int A(x)\,dx$
$a\,x^n$	$\frac{a}{n+1}\,x^{n+1}$
e^x	e^x
$\ln ax$	$x\ln ax - x$
$\sin ax$	$-\frac{1}{a}\cos ax$
$\cos ax$	$\frac{1}{a}\sin ax$
$\tan ax$	$-\frac{1}{a}\ln(\cos ax)$
$\operatorname{cosec} ax$	$\frac{1}{a}\ln(\operatorname{cosec} ax - \cot ax)$
$\sec ax$	$\frac{1}{a}\ln(\sec ax + \tan ax)$
$\cot ax$	$\frac{1}{a}\ln(\sin ax)$

Bibliography

Books

Arya, A. P. (1979) *Introductory College Physics*, McMillan Publishing Co., Inc., New York

Beiser, A. (1973) *Concepts of Modern Physics*, Mc-graw Hill, Kogakusha, Ltd., Japan

Born, M. and E. Wolf (1975) *Principles of optics* Pergamon Press, Oxford

Born, M. and E. Wolf (1980) *Principles of Optics: Electromagnetic theory of prorogation, interference and diffraction of light*, Cambridge University Press, UK

Ditchburn, R.W. (1976) *Light*, Academic Press, London

Fishbane, P. M., S. Gasiorowitz and S. T. Thorton (1996) *Physics for Scientists and Engineers*, Prentice Hall

Giancoli, D. C. (2000) *Physics for Scientists and Engineers*, Prentice Hall

Griffiths, D. J. (1989) *Introduction to Electrodynamics*, Prentice Hall International, Inc., New jersey

Ingram, D.J.E. (1973) *Radiation and quantum physics*, Clarendon Press, Oxford

Jenkins, F. A. and H. E. White, H. E. (1981) *Fundamentals of Optics*, Mc-Graw Hill Book Company, London

Landau, L. D. and E. M. Liftshitz, E M. (1960)*Electrodynamics of continuous media*, (Oxford:Pergamon Press)

Lipson, H.S. (ed.) (1972) *Optical transforms*, Academic Press, New York

Longhurst, R.S. (1973) *Geometrical and physical optics*, Longmans, London

Loudon, R. (1973) *Quantum theory of light*, Clarendon Press, Oxford

Midwinter, J.E. (1979) *Optical fibres for transmission*, John Wiley, New York

Nkoma, J. S. (2018) *Introduction to Basic Concepts for Engineers and Scientists: Electromagnetic, Quantum, Statistical and Relativistic Concepts*, Mkuki na Nyota, Dar es Salaam

Ohanian, H. C. (1985) *Physics*, W W Norton & Company

Robinson, F.N.H. (1973) *Electromagnetism*, Clarendon Press, Oxford

Sears, F. W., M. W. Zemansky and H. D. Young (1985) *College Physics*, Addison-Wesley

Serway, R. A. (1990) *Physics for Scientists and Engineers with Modern Physics*, Saunders

Smith, C. J. (1943) *Intermediate Physics*, Edward Arnold and Co., London

Stratton, J.A. (1941) *Electromagnetic theory*, (New York: McGraw-Hill)

Svelto, O. (1976) *Principles of lasers*, (trans. D.C. Hanna). Heyden, London

Tipler, P. A. (1999) *Physics for Scientists and Engineers*, W H Freeman and Company/ Worth Publishers, New York

Welford, W.T. (1962) *Geometrical optics; optical instrumentation*, North-Holland, Amsterdam

Welford, W.T. (1974) *Aberrations of the symmetrical optical system*, Academic Press, London

Welford, W.T. (1981) *Optics*, Oxford University Press

Wolfson, R and J. M. Pasachoff (1999) *Physics*, Addison-Wesley

Young, H D and R. A. Freedman (2000) *University Physics*, Addison-Wesley

Articles

Crease, R. P. (September 2002) *"The most beautiful experiment"*, Physics World, **15**, Number 8, 19, and also *The Editorial Comment: "The double-slit experiment"* page 15

Jain, P. K. and L. K. Sharma (1988) *"The Physics of Blackbody Radiation: A Review"*, J Appl. Sc. in Southern Afr., **4**, 80 - 101

Nkoma, J. S. and G. Ekosse (1999) *"X-ray Diffraction Study of Chalcopyrite $CuFeS_2$, Pentlandite $(Fe,Ni)_9S_8$ and Pyrrhotite $Fe_{1-x}S$ obtained from Cu-Ni orebodies"*, J. Phys. Condens. Matter **11** 121 - 128

Provostaya, F. and P. Desanis (1850) *Ann. Chim. Phys.*, **30**, 159

Index

Printed in the United States
By Bookmasters